PE EXAM PREP

CHEMICAL ENGINEERING

PE SAMPLE EXAM

Second Edition

Rajaram K. Prabhudesai, PhD, PE Chem. Eng.

President: Roy Lipner
Vice President & General Manager: David Dufresne
Vice President of Product Development and Publishing: Evan M. Butterfield
Editorial Project Manager: Laurie McGuire
Director of Production: Daniel Frey
Production Editor: Caitlin Ostrow
Creative Director: Lucy Jenkins

Published by Kaplan AEC Education
30 South Wacker Drive
Chicago, IL 60606-7481
(312) 836-4400
www.kaplanaecengineering.com

Printed in the United States of America.

08 09 10 10 9 8 7 6 5 4 3 2 1

CONTENTS

Introduction

OUTLINE

HOW TO USE THIS BOOK

Chemical Engineering PE Sample Exam gives you an opportunity to simulate the experience of taking the Principles and Practice of Engineering exam in chemical engineering. It is hoped that this experience will make you feel more relaxed and prepared on the day of the actual exam. The problems and solutions covered in this book will also give you a good review of concepts and analytical techniques that are likely to appear on the exam. A good approach to using this book for optimal exam preparation follows:

1. Set aside a 4-hour block of time to answer the 40 questions in Chapter 1. Don't look at the solutions until you have answered all the questions. Use the reference texts you plan to bring to the actual exam, as needed, to help you choose the correct answer to each question. Appendix C provides fill-in answer sheets that model the ones you will see in the actual exam.

2. Set aside another 4-hour block of time to answer the 40 questions in Chapter 2. Follow the same procedures as for Chapter 1. To closely approximate the exam experience, you may want to tackle Chapters 1 and 2 consecutively on the same day, with just an hour-long break between them. This is the format of the actual exam.

3. When you have completed the exams, turn to Chapter 4 to review the correct answers and detailed solutions. Use the solution and topic summary tables to make note of which topic areas gave you the most difficulty—these are areas you want to focus on further as you review for the exam.

4. Turn to the additional review questions in Chapter 3 for further practice in the major topic areas of the exam. Again, it's a good idea to solve the problem independently before looking at the solution. If you answer a question incorrectly, the detailed solution provided in Chapter 5 should help you understand the correct approach for similar problems in the future.

Some of the questions in this book may be a bit more difficult or lengthy to solve than those on the actual exam, but their technical content is worth the effort. It is better to be overly prepared.

For a more thorough review for the exam, you may want to use this book in combination with its companion texts: *Chemical Engineering PE License Review* and *Chemical Engineering PE Problems & Solutions. Chemical Engineering PE License Review* provides a conceptual review of key terms and equations, analytical methods, and design considerations. *Chemical Engineering PE Problems & Solutions* provides an opportunity for extensive practice solving problems; it is organized in the same topic order as the *License Review* book so that you can refer back and forth between problems and concepts as needed. Together, these three books provide a comprehensive review for the exam.

BECOMING A PROFESSIONAL ENGINEER

There are four distinct steps to achieving registration as a Professional Engineer: (1) education; (2) passing the Fundamentals of Engineering/Engineer-in-Training (FE/EIT) exam; (3) professional experience; and (4) passing the Professional Engineer (PE) exam, more formally known as the Principles and Practice of Engineering Exam. These steps are described in the following sections.

Education

The obvious appropriate education is a B.S. degree in chemical engineering from an accredited college or university. This is not an absolute requirement. Alternative, but less acceptable, education is a B.S. degree in a field other than chemical engineering, a degree from a non-accredited institution, or 4 years of education but no degree.

Fundamentals of Engineering/Engineer-in-Training (FE/EIT) Exam

Most candidates for PE registration are required to take and pass this 8-hour multiple-choice examination. Although different states call it by different names (e.g., Fundamentals of Engineering, Engineer-in-Training, or Intern Engineer), the exam is the same in all states. It is prepared and graded by the National Council of Examiners for Engineering and Surveying (NCEES). Review materials for this exam are found in other books published by Kaplan AEC, such as *Fundamentals of Engineering: FE/EIT Exam Preparation.*

Experience

Several years of acceptable experience are typically required before an engineer is permitted to take the PE exam. Both the length and character of the experience will be examined.

Professional Engineer Exam

The second national exam is called Principles and Practice of Engineering by the NCEES, but almost everyone else calls it the Professional Engineer, or PE, exam. All U.S. states, the District of Columbia, Guam, and Puerto Rico use the same NCEES exam.

CHEMICAL ENGINEERING PROFESSIONAL ENGINEER EXAM

Laws regulating the practice of engineering are adopted to protect the public from incompetent practitioners. Most states require engineers who work on projects involving public safety to be registered, or to work under the supervision of a registered professional engineer. In addition, many private companies encourage or require engineers in their employ to pursue registration as a matter of professional development. Engineers in private practice and those who wish to consult or serve as expert witnesses typically also must be registered. There is no national registration law; registration is based on individual state laws and is administered by boards of registration in each of the states. You can find a list of contact information for and links to the various state boards of registration at the Kaplan AEC Web site: *www.kaplanaecengineering.com.* This list also shows the exam registration deadline for each state.

Examination Development

Initially, the states wrote their own examinations, but beginning in 1966 the NCEES took over the task for some of the states. Now the NCEES exams are used by all states. This makes it easier for an engineer to move from one state to another and achieve registration in the new state.

The development of the engineering exams is the responsibility of the NCEES Committee on Examinations for Professional Engineers. The committee is composed of representatives from industry, consulting, and education, as well as consultants and subject matter experts. The starting point for the exam is an engineering task analysis survey that the NCEES does at roughly 5- to 10-year intervals. People in industry, consulting, and education are surveyed to determine what chemical engineers do and what knowledge is needed to do it. From this data, the NCEES develops what they call a "matrix of knowledge" that forms the basis for the exam structure described in the next section.

The actual exam questions are prepared by the NCEES committee members, subject matter experts, and other volunteers. All participants must hold professional registration. Using workshop meetings and correspondence by mail, the questions are written and circulated for review. Although based on an understanding of engineering fundamentals, the problems require the application of practical professional judgment and insight.

Examination Structure

The exam consists of 80 multiple-choice questions, which cover the following areas of chemical engineering (the relative exam weight for each topic is shown in parentheses):

- Mass/energy balances and thermodynamics (24%)
- Fluids (17%)
- Heat transfer (16%)
- Mass transfer (13%)
- Kinetics (11%)
- Plant design and operations (19%)

For more information on the topics, subtopics and their relative weights on the breadth and depth portions, visit the NCEES Web site at *www.ncees.org*.

The exam is given over two 4-hour sessions, with 40 questions in each session. All questions are multiple choice, with four answer choices.

Exam Dates

The NCEES prepares Professional Engineer exams for administration on one Friday in April and one Friday in October of each year. Some state boards administer the exam twice a year in their state, while others offer the exam once a year. The scheduled exam dates for the next ten years can be found on the NCEES Web site (*www.ncees.org/exams/schedules/*).

People seeking to take a particular exam must apply to their state board of registration several months in advance.

Exam Procedure

Before the morning 4-hour session begins, the proctors pass out an exam booklet and a solutions pamphlet to each examinee. The solutions pamphlet contains grid sheets on right-hand pages. Only work on these grid sheets will be graded. The left-hand pages are blank and are for use as scratch paper. The scratch work will *not* be considered in the scoring.

The proctors also will provide each examinee with a mechanical pencil for use in recording answers; this is the only writing instrument allowed. Do not bring your own pencil lead or eraser. If you need an additional pencil during the exam, a proctor will supply one.

If you finish more than 15 minutes early, you may turn in the booklets and leave. If you finish in the last 15 minutes, however, you must remain through the end of the hour to ensure a quiet environment for those still working, and to ensure an orderly collection of materials.

The afternoon session will begin following a 1-hour lunch break. The afternoon exam booklet will be distributed along with an answer sheet.

Exam-Taking Suggestions

Give yourself time to prepare for the exam in a calm and unhurried way. Many candidates like to begin several months before the actual exam. Target a number of hours per day or week that you will study, and reserve blocks of time for doing so. Creating a review schedule on a topic-by-topic basis is a good idea. Remember to allow time for both reviewing concepts and solving practice problems.

In addition to review work that you do on your own, you may want to join a study group or take a review course. A group study environment may help you stay committed to a study plan and schedule. Group members can create additional practice problems for one another and share tips and tricks.

You may want to prioritize the time you spend reviewing specific topics according to their relative weight on the exam, as identified by the NCEES, or according to your areas of relative strength and weakness. There may be an exam topic that you have little or no exposure to. This would be a good area to focus on, time permitting, provided that you feel strong in other areas.

People familiar with the psychology of exam-taking have several suggestions for exam candidates:

1. Passing a competency exam involves two skills. One is the skill of illustrating your knowledge. The other is the skill of exam-taking. The first may be enhanced by a systematic review of the technical material. Exam-taking skills, on the other hand, may be improved by practicing with problems that are presented in a format similar to the exam format.

2. Because there is no penalty for guessing on the multiple-choice problems, you should answer all of the problems. Even when you are going to guess, use a logical approach. Attempt to first eliminate one or two of the four alternatives. If you can do this, the chance of selecting a correct answer obviously improves, from 1 in 4 to 1 in 3 or 1 in 2.

3. Plan ahead with a strategy. Which is your strongest area? Can you expect to see several problems in this area? What about your second-strongest area? What is your weakest area?

4. Plan ahead with a time allocation. Compute how much time you will allow for each of the major subject areas. You might allocate a little less time per problem for those areas in which you are most proficient, leaving a little more time in subjects that are more difficult for you. Your plan should include a block of time reserved for especially difficult problems, for checking your scoring sheet, and finally for making last-minute guesses on any problems you did not work. Your strategy may also include time allotments for two passes through the exam: the first to work all problems for which answers are obvious to you; the second to return to the more complex, time-consuming problems and the ones at which you may need to guess. A plan mapping out how you will spend your time gives you the confidence of being in control and keeps you from making the serious mistake of misallocating time in the exam.

5. Read all four multiple-choice answers before making a selection. An answer in a multiple-choice question is sometimes a plausible decoy—not the best answer.

6. Do not change an answer unless you are absolutely certain you have made a mistake. Your first reaction is likely to be correct.

7. Do not sit next to a friend, a window, or other potential distractions.

Exam Day Preparations

The exam day will be a stressful and tiring one. This will be no day to have unpleasant surprises. For this reason, we suggest that an advance visit be made to the examination site. Try to determine such items as the following:

1. How much time should I allow for travel to the exam on that day? Plan to arrive about 15 minutes early. This way you will have ample time, but not too much time. Arriving too early and mingling with others who are anxious will increase your anxiety and nervousness.
2. Where will I park?
3. How does the exam site look? Will I have ample workspace? Where will I stack my reference materials? Will it be overly bright (bring sunglasses) or cold (bring a sweater), or noisy (bring earplugs)? Would a cushion make the chair more comfortable?
4. Where are the drinking fountain and lavatory facilities?
5. What about food? Should I bring a snack for energy during the exam? A bag lunch during the break probably makes sense.

What to Take to the Exam

The NCEES guidelines allow you to bring the following reference materials and aids into the examination room for your personal use only:

1. Handbooks and textbooks, including the applicable design standards.
2. Bound reference materials, provided the materials remain bound during the entire examination. The NCEES defines "bound" as books or materials fastened securely in their covers by fasteners that penetrate all the papers. Examples are ring binders, spiral binders and notebooks, plastic snap binders, binders with brads or screw posts, and so on.
3. A battery-operated, silent, non-printing, non-communicating calculator from the NCEES list of approved calculators. For the most current list, see the NCEES Web site (*www.ncees.org*). You must also determine whether your state permits preprogrammed calculators. Bring extra batteries for your calculator just in case; many people feel that bringing a second calculator is also a very good idea.

At one time, the NCEES prohibited bringing "review publications directed principally toward sample questions and their solutions" into the exam room. This led to restrictions against bringing some kinds of publications to the exam. *State boards may adopt the NCEES guidelines, or may adopt either more or less restrictive rules.* Thus an important step in preparing for the exam is to know what will—and will not—be permitted at your exam location. We recommend that you obtain a written copy of your state's policy for the specific exam you will be taking. Occasionally there has been confusion at individual examination sites, so a copy of the exact applicable policy will not only enable you to carefully and correctly prepare your materials, but it will also ensure that the exam proctors allow all proper materials that you bring to the exam.

As a general rule, we recommend that you plan well in advance what books and other materials you want to take to the exam. Obtain them promptly so that you can use the same materials in your review that you will use in the exam.

License Review Books

The review books you use to prepare for the exam are good choices to bring to the exam itself. After weeks or months of studying, you will be very familiar with their organization and content, so you'll be able to quickly locate the material you want to reference during the exam. Keep in mind the caveat just discussed—some state boards will not permit you to bring review books that consist largely of sample questions and answers into the exam room.

Textbooks

If you still have your college or university textbooks, they are the ones you should use in the exam, unless they are too out of date. Because of their familiar notation, these books will be like old friends.

Bound Reference Materials

The NCEES guidelines suggest that you can take to the exam any reference materials you wish, so long as you prepare them properly. You could, for example, prepare several volumes of bound reference materials with each volume covering a particular category of problem. Use tabs so that specific material can be located quickly. If you do a careful and systematic review, and prepare a lot of well-organized materials, you just may find that you are so well prepared that you will not have left anything of value at home.

Other Items

In addition to the reference materials just mentioned, you should consider bringing the following to the exam:

- Clock—You must have a time plan and a clock or wristwatch.
- Exam assignment documents—Take along the letter assigning you to the exam at the specified location. To prove you are the correct person, also bring at least one photo ID (such as a driver's license).
- Items suggested by advance visit—If you visit the exam site, you will probably discover an item or two that you need to add to your list.
- Clothing—Plan to wear comfortable clothes. You will probably do better if you are slightly cool.
- Box for everything—You need to be able to carry all your materials to the exam and have them conveniently organized at your side. A cardboard box is probably the answer.

Examination Scoring and Results

The questions are machine-scored by scanning. The answers sheets are checked for errors by computer. Marking two answers to a question, for example, will be detected and no credit will be given.

Your state board will notify you whether you have passed or failed roughly three months after the exam. Candidates who do not pass the exam the first time may take it again. If you do not pass you will receive a report listing the percentages of questions you answered correctly for each topic area. This information can help focus the review efforts of candidates who need to retake the exam.

The PE exam is challenging, but analysis of previous pass rates shows that the majority of candidates do pass it the first time. By reviewing appropriate concepts and practicing with exam-style problems, you can be in that majority. Good luck!

CHAPTER 1

Morning Exam

1.1 A liquid fuel has the following composition:

Component	Wt %	Atomic Weight
C	87.94	12.01
H	10.57	1.01
S	0.84	32.0
O	0.65	16.0

This fuel is burned in a boiler with 20% excess air. Assuming complete combustion, the actual oxygen used (lb per 100 lb of fuel burned) is nearly:

a. 483
b. 318
c. 382
d. 428

1.2 A natural gas has the following composition by volume:

CH_4, 97.3%; N_2, 2.3%; and CO_2, 0.4%

It is burned in a furnace. The Orsat analysis of the flue gases shows the following:

CO_2, 8.8%; O_2, 5.26%; and N_2, 85.94%

The percent excess air supplied to the furnace is nearly:

a. 20%
b. 15%
c. 25%
d. 30%

1.3 Calcium carbide reacts with water according to the following reaction:

$$CaC_2 + 2H_2O \rightarrow C_2H_2 + Ca(OH)_2$$

The molecular weights of the reactants and products in the above reaction are as follows:

Component	MW
CaC_2	64.1
H_2O	18.02
C_2H_2	26.04
$Ca(OH)_2$	74.1

An acetylene lamp uses 3 ft^3 of gas measured at a temperature of 75°F and a pressure of 1 atm per hour. The amount of calcium carbide (lb) needed to yield enough acetylene to fuel the lamp for 24 hours is nearly:

a. 11.82
b. 10.10
c. 14.60
d. 9.36

1.4 A plant makes very high-grade lime by calcining pure precipitated $CaCO_3$ in rotary kilns. In one kiln, the reaction goes to 95% completion. The calcium carbonate decomposes according to the following reaction:

$$CaCO_3 \rightarrow CaO + CO_2$$

The molecular weights for the compounds are listed in the following table.

Component	MW
$CaCO_3$	100
CaO	56
CO_2	44

The yield, in terms of lb of CO_2 produced per lb of calcium carbonate, is nearly:

a. 0.242
b. 0.418
c. 0.324
d. 0.366

1.5 The composition of a natural gas is 90% CH_4 and 10% N_2 by volume. It is burned under a boiler, and the CO_2 is scrubbed out from the cooled dry flue gases for the production of dry ice. The Orsat analysis of the gas leaving the scrubber shows 1.1% CO_2, 5% O_2, and 93.9% N_2. A schematic of the process is shown in Exhibit 1.5.

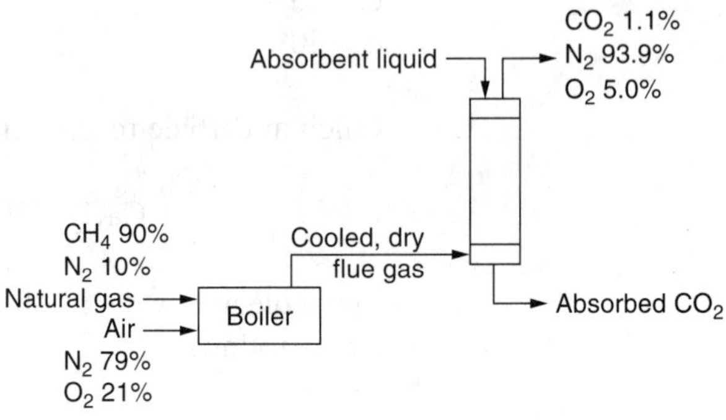

Exhibit 1.5 Sketch of the process for Problem 1.5

Methane burns according to the following reaction:

$$CH_4 + 2O_2 \rightarrow CO_2 + 2H_2O$$

In this process, 25% excess air is used. Assume complete combustion of the methane in the boiler furnace. The percent CO_2 absorbed in the scrubber is nearly:

a. 88.9
b. 95.2
c. 76.4
d. 85.7

1.6 Saturated steam at 150°C is fed to a condenser at a flow rate of 2500 kg/min. The quality of vapor leaving the condenser is 50%. (See Exhibit 1.6.)

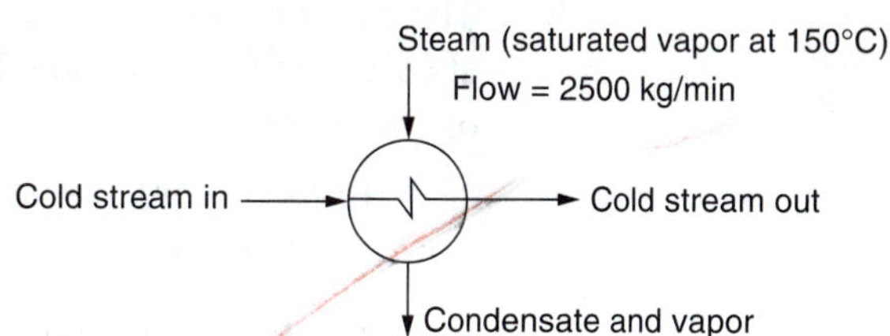

Exhibit 1.6 Process sketch for Problem 1.6

The amount of heat removed from the condenser per minute is nearly:

a. 3.2 MMBtu/min
b. 3×10^6 kJ/min
c. 2.5 MMBtu/min
d. 2.3×10^6 kJ/min

1.7 An endothermic reaction is carried out in a jacketed reaction vessel with an initial charge (average specific heat = 0.8 Btu/lb·°F) of 500 lb, consisting only of reactants. The feed charge temperature is 68°F. The reaction is carried out for 2 hours. Heat is supplied to the reactor by condensing saturated steam at 450°F in the jacket. The reaction mass absorbs heat at 991 Btu/lb of charge during the reaction, and heat losses are 8000 Btu/h of reaction time. At the end of 2 hours, the temperature of the reaction mass is 212°F. The reactor system is shown in Exhibit 1.7.

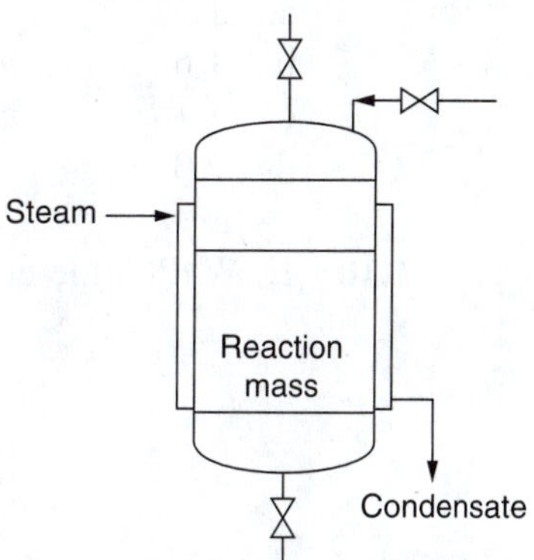

Exhibit 1.7 Process sketch for Problem 1.7

Enthalpy data for steam, taken from the steam tables, is given below:

Temperature, °F	Enthalpy of Liquid, Btu/lb	Enthalpy of Vapor, Btu/lb	Heat of Vaporization λ_S, Btu/lb
440	419.0	1204.4	785.4
460	441.5	1204.8	763.3

Under these conditions, the steam usage (lb/batch) is nearly:
a. 735
b. 824
c. 916
d. 863

1.8 The heat of combustion of gaseous isopentane [2-methyl butane (C_5H_{12})] at 25°C and 1 atm is –843.216 kcal/g mol based on water as liquid. The heats of formation of the combustion products are as follows:

CO_2(g) $\Delta H_f = -94.05$ kcal/g mol at 25°C and 1 atm
H_2O(l) $\Delta H_f = -68.316$ kcal/g mol at 25°C and 1 atm

The heat of formation [kcal/g mol] of isopentane at 25°C and 1 atm is nearly:
a. –88.02
b. –36.93
c. –84.32
d. –63.10

1.9 Air is compressed from 1 atm and 0°F ($\hat{H}_I = 210.27$ Btu/lb) to 10 atm and 40°F ($\hat{H}_O = 218.87$ Btu/lb). The exit velocity of air from the compressor is 200 ft/s. Inlet air velocity is negligible. These data are shown in Exhibit 1.9.

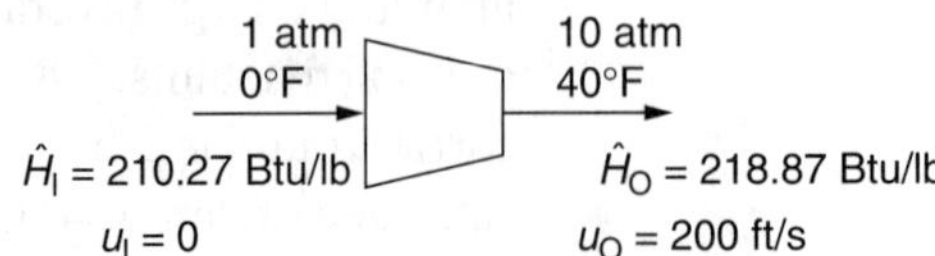

Exhibit 1.9 Process sketch for Problem 1.9

If the load is 5 lb/min of air and the efficiency of the compression is 70%, then the break horsepower (bhp) required by the compressor is:
a. 1.3
b. 1.6
c. 1.1
d. 2.0

1.10 H. W. Prengle et al. report the following data for n-butane:

t, °F	*P*, atm	V_l, ft³/lb	V_g, ft³/lb
260	24.662	0.0393	0.222
270	27.134	0.0408	0.192
280	29.785	0.0429	0.165
290	32.624	0.0458	0.138
305.56	37.47	0.0712	0.0712

The latent heat of vaporization of n-butane (Btu/lb) at 280°F is nearly:

a. 102
b. 312
c. 67.5
d. 220

1.11 A solution of specific gravity 1.84 is being discharged from a tank into the atmosphere. The level of the liquid in the tank is 20 ft above the centerline of the exit pipe. Losses in the pipe due to friction and contraction amount to 12-ft head of solution.

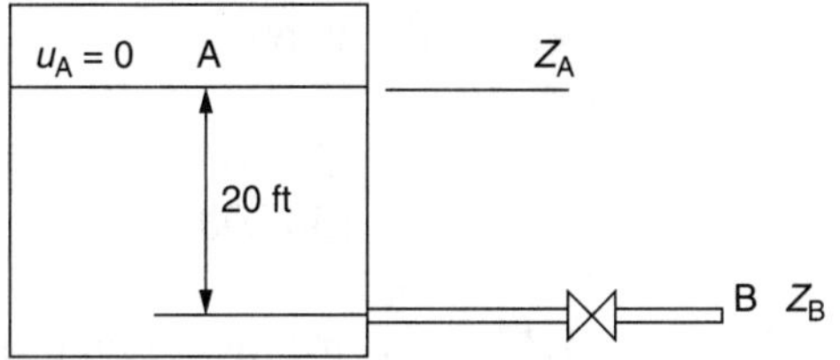

Exhibit 1.11 Process sketch for Problem 1.11

Under these conditions, the discharge velocity (ft/s) from the pipe is nearly:

a. 35.2
b. 7.4
c. 22.7
d. 27.8

1.12 A manufacturer lists the following C_v data for his line of control valves.

Control Valve Size, in.	1	1.5	2	3	4	6	8
Valve Coefficient C_v	45	125	165	350	775	1000	2000

The service conditions for a certain process are as follows:

Fluid	**Aqueous Solution**
Flow rate (gpm), min/max	125/1000
Flow rate (normal) gpm at F.T.	400
Pressure (psig), max in/normal out	50/25
ΔP (psi), max/design	30/4
Temperature (°F), max/normal	100/75
Specific gravity at 60°F/at F.T.	(NA)/1.2

The size (in.) of a suitable control valve for this service will be:

a. 3
b. 6
c. 4
d. 8

1.13 The relative roughness for an 8-in., Schedule 80 commercial steel pipe is nearly:
a. 1.92×10^{-4}
b. 1.50×10^{-4}
c. 2.36×10^{-4}
d. 2.03×10^{-4}

1.14 Oil is flowing through a 6-in., Schedule 40 (inside diameter 0.5054 ft and cross-sectional area 0.2006 ft^2) steel pipe at a rate of 500 gpm. At the flow temperature, the specific gravity of oil is 0.9 while its viscosity is 4.9 lb/h · ft. The Reynolds number of the oil flowing through the pipe is nearly:
a. 38,600
b. 47,900
c. 85,200
d. 138,200

1.15 A centrifugal pump draws a solution (specific gravity = 1.605) from a storage tank with a large cross section through a standard 3-in. Schedule 40 steel pipe and delivers it into a 2-in. Schedule 40 pipe.

The liquid velocity in the suction line is 3 ft/s. The level of solution in the tank is maintained at 10 ft from the centerline of the pump. The suction-head loss from the storage tank to the pump is 1.5 ft of head of solution, and the pump develops a head of 43.2 ft of solution. The pressure (psig) shown by a gage at the pump discharge will be nearly:
a. 20.8
b. 35.5
c. 50.2
d. 43.2

1.16 Air is flowing in a duct 20-in. i.d. (flow area = 2.182 ft^2) at 220°F and 750 mm Hg pressure. A pitot tube positioned at the center of the cross section of the pipe shows a manometer reading of 4-in. water column. The pitot tube coefficient is 0.98. The viscosity of air is 0.022 cP. The air flow rate (scfm at 60°F and 1 atm) through the duct is nearly:
a. 16,022
b. 12,093
c. 15,000
d. 10,500

1.17 Water is being pumped from an open tank and discharged to another point as shown in the figure below. (*Not* all of the fittings are shown in Exhibit 1.17.)

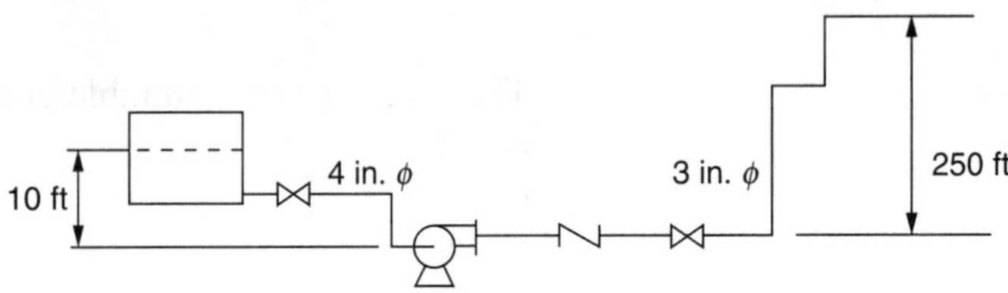

Exhibit 1.17 Process sketch for Problem 1.17

The following data are available:

	Suction Side	Discharge Line
Pressure above liquid surface in the tank, psia	14.7	
Discharge pressure, psia		14.7
Velocity, ft/s	3.02	5.21
Length of piping, ft	12	250.00
Height of liquid surface from centerline of pump, ft	10	
Discharge point elevation from centerline of pump, ft		250.00
Sum of resistance coefficients of fittings, ΣK	3.22	8.77

Other data include the following:

Density of liquid water = 62.4 lb/ft^3
Inside diameter of discharge piping = 0.2557 ft
Inside diameter of suction piping = 0.3355 ft
Suction side pressure drop = 0.24 psi
Fanning friction factor for the discharge line = 0.0049
Neglect the velocity head on either side of pump.

If the pump efficiency is 70%, the brake horsepower (bhp) of the pump is nearly:

a. 13
b. 15
c. 11
d. 9

1.18 The composite walls of a furnace are made of 9-in. kaolin firebrick, 4-in. insulating brick, and 8-in. building brick. The mean thermal conductivities of the materials are as follows:

Material	Thermal Conductivity, Btu/(h·ft^2·°F/ft)
Kaolin fire brick	0.68
Insulating brick	0.15
Building brick	0.40

When the inside surface of the furnace wall is maintained at 2000°F, the outside surface temperature of the wall was found to be 180°F. The interface temperature (°F) between the kaolin and insulating bricks will most nearly be:

a. 1592
b. 1500
c. 1700
d. 1800

1.19 The following Dittus-Boelter equation can be used to calculate tube-side convection heat-transfer coefficients under certain conditions.

$$\frac{h_i D_i}{k_b} = \left(\frac{D_i G}{\mu_b}\right)^{0.8} \left(\frac{c_P \mu}{k}\right)_b^{0.3}$$

The correct tube-side conditions under which the above equation can be used are:
a. Moderate Δt values, heating, and turbulent flow
b. Moderate Δt values, cooling, and turbulent flow
c. Any Δt values and turbulent flow
d. Any laminar, turbulent, or transition flow

1.20 A double-pipe countercurrent heat exchanger (outside surface area = 200 ft^2) has reached its design capacity of 490,000 Btu/h and is scheduled to be shut down for cleaning. The data for the exchanger are as follows:

	Shell Side	Tube Side
Temperature of hot fluid, °F, in	350	
Temperature of hot fluid, °F, out	450	
Temperature of cold fluid, °F, in		300
Temperature of cold fluid, °F, out		310
Heat transfer coefficient, Btu/(h·ft^2·°F)	38.4	300
Tubes are 1-in. o.d. 18 BWG		

The dirty overall coefficient of heat transfer [Btu/(h·ft^2·°F)] for the heat exchanger is nearly:
a. 50
b. 40
c. 28
d. 35

1.21 The 45,000 lb/h of kerosene (c_P = 0.605 Btu/lb·°F) leaving the bottom of a distillation column at 390°F is cooled to 200°F using cold crude stream (specific heat = 0.49 Btu/lb·°F and flow = 150,810 lb/h) at 100°F. The corrected mean temperature difference (°F) that should be used in the design of a one-shell, four-tube-passes heat exchanger for this service is nearly:
a. 138
b. 152
c. 145
d. 160

1.22 In a surface condenser, steam is condensing on the shell side and a liquid is being heated on the tube side. The tubes are 18-gage and 1-in. o.d. At the end of 1 year of operation, the film heat-transfer coefficients and the dirt factors are as follows:

Steam film heat-transfer coefficient on the shell side = 2000 Btu/(h·ft^2·°F)
Dirt deposit coefficient on the shell side = 2000 Btu/(h·ft^2·°F)
Dirt deposit coefficient on the tube side = 2000 Btu/(h·ft^2·°F)
Liquid film heat-transfer coefficient on the tube side = 1800 Btu/(h·ft^2·°F)

The thermal conductivity of the tube material = 63 Btu/(h·ft^2·°F/ft).

After thoroughly cleaning the tubes on both sides, the percent increase in the overall heat-transfer coefficient will be nearly:
a. 15%
b. 20%
c. 30%
d. 24.5%

1.23 A bare horizontal steel pipe (o.d. 115 mm) is carrying steam at 400 K. The surrounding air is at 21.1°C. The convective heat transfer coefficient h_c is given by the following equation:

$$h_c = 1.0813\left(\frac{\Delta t}{d_O}\right)^{0.25}$$

where

h_c = convection heat transfer coefficient, W/m^2·K
d_O = outside diameter, m
t = temperature difference between pipe surface and the surrounding air, K

The emissivity of the pipe surface = 0.8. Therefore, the heat loss per meter length of pipe (kW/m) is nearly:

a. 211
b. 507
c. 359
d. 289

1.24 Oxygen (component A) is diffusing through stagnant carbon monoxide (B) under steady state conditions. Carbon monoxide is non-diffusing. The total pressure is 1 atm, and the temperature is 32°F. The partial pressures of oxygen at two parallel planes 0.12 in. apart are 0.13 and 0.065 atm respectively. The diffusivity of oxygen in the mixture is 0.7244 ft^2/h. The rate of diffusion of oxygen [(lb mol)/(h·ft^2)] through the gas mixture is nearly:

a. 0.0145
b. 0.042
c. 0.031
d. 0.039

1.25 i-Butanol forms a minimum boiling-point azeotrope. The *t-x* diagram for this system is shown in Exhibit 1.25.

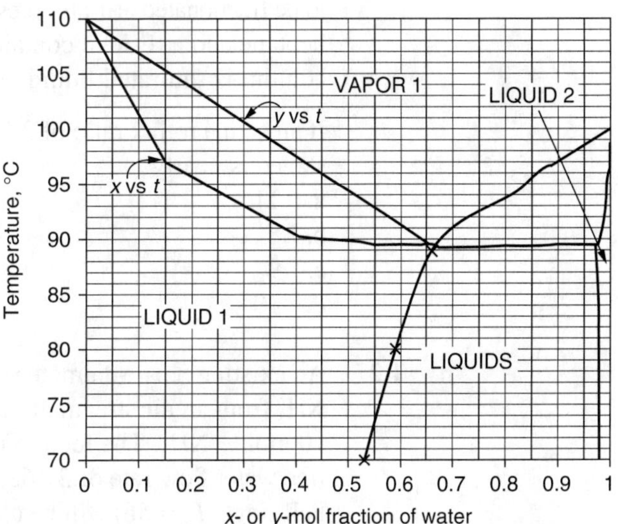

Exhibit 1.25 A *t-x* diagram for an i-Butanol–water system (Problem 1.25)

An aqueous mixture of i-butanol and water contains 95 wt % water. The bubble point (°C) of this liquid is most nearly:

a. 94
b. 89.8
c. 100
d. 97

1.26 Exhibit 1.26 is an equilibrium diagram showing the vapor-liquid compositions of heptane at 1 atm pressure for the system heptane–ethyl benzene.

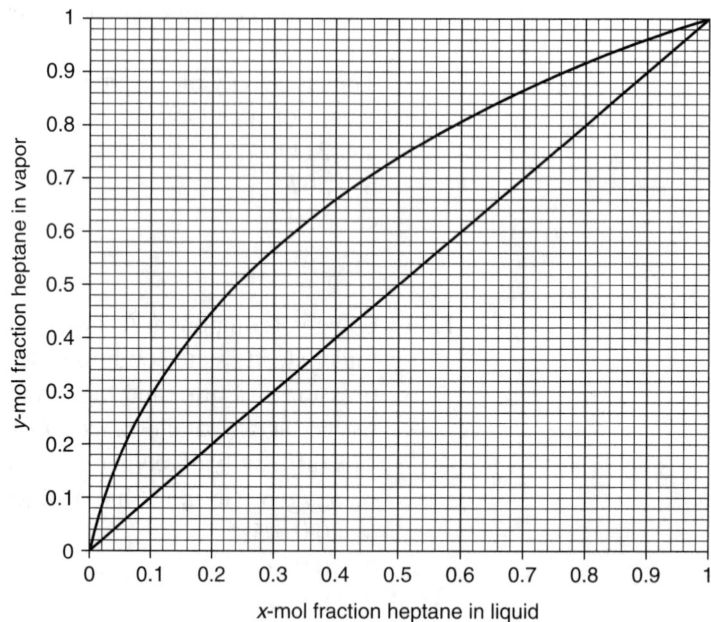

Exhibit 1.26 Equilibrium diagram for heptane–ethyl benzene system for Problem 1.26

A feed mixture containing 40 mol % heptane and 60 mol % ethyl benzene is to be fractionated at 1 atm pressure to produce a distillate containing 97 mol % heptane and a residue containing 98 mol % ethyl benzene. The feed to the column is saturated liquid, and the reflux is at its bubble point.

If an actual reflux ratio of 2.5 is used, the theoretical number of equilibrium stages is nearly:

a. 11
b. 8
c. 9
d. 10

1.27 An existing 5.5-ft-diameter tower is going to be used to absorb 99% of the NH_3 from an air stream at 16 psia and a temperature of 80°F that contains 10 mol % NH_3. The feed to the tower will be 1000 lb mol/h. An ammonia-free water flow rate of 37,620 lb/h (2090 lb mol/h) will be used. Also, 1-in. Pall rings ($F_p = 56$) will be used as packing. Exhibit 1.27 depicts the known process conditions at the two ends of the tower.

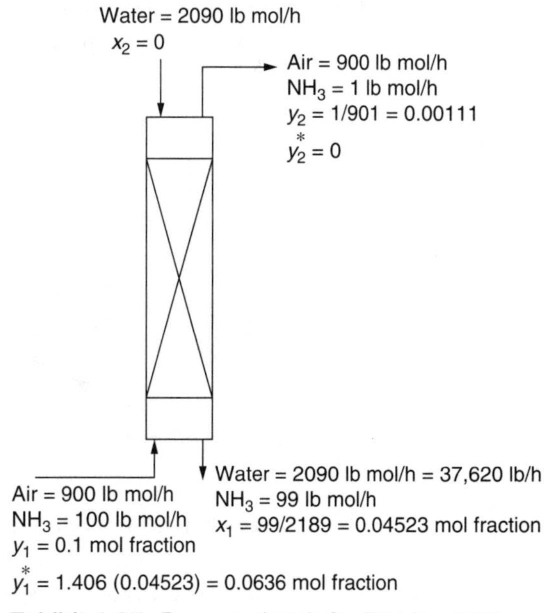

Exhibit 1.27 Process sketch for Problem 1.27

The equilibrium relationship over the range of concentrations involved can be represented by the following equation:

$$y = 1.406x$$

where

y = mol fraction of NH_3 in vapor
x = mol fraction of NH_3 in liquid

The temperature in the tower will be maintained constant at 80°F by means of cooling coils.

If the overall mass transfer coefficient is 16.2 lb mol/(h·ft^3·mol fraction), the packed height (ft) needed is most nearly:

a. 20
b. 25.5
c. 30
d. 36

1.28 A packed tower is to be designed to treat 30,000 ft^3 of gas per hour to remove NH_3 from it. The ammonia content of the gas is 5% by volume. Ammonia-free water will be used as an absorbent. The temperature is 68°F, and the pressure is 15 psia. The ratio of the liquid flow rate to the gas flow rate is 1. Also, 1.5-in. ceramic Intalox saddle rings (F_p = 52) will be used as packing.

If the column is to be designed for a pressure drop of 0.5 in. of H_2O per foot of packing, the diameter (in.) of the column will be nearly:

a. 18
b. 30
c. 24
d. 36

1.29 A reversible reaction of the type $A \rightleftarrows B$ was studied with $C_{A0} = 0.8$ lb mol/ft^3 and $C_{B0} = 0.5$ lb mol/ft^3. The equilibrium conversion X_{AE} is 0.35. The conversion data taken as a function of time are plotted in Exhibit 1.29.

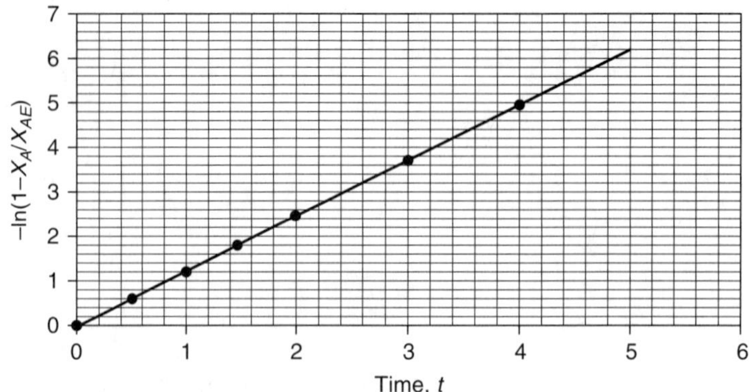

Exhibit 1.29 Plot of the conversion data for Problem 1.29

The forward-reaction specific rate constant [h^{-1}] is nearly:
a. 0.93
b. 0.743
c. 0.66
d. 0.84

1.30 Specific rate constants were experimentally determined for a reaction at various temperatures and are listed below:

T, K	303	313	323	333	343
k, $\frac{\text{liters}}{\text{g mol} \cdot \text{K}}$	0.5	1.1	2.2	4.0	6.0

These data are plotted as ln k versus $1/T$ in Exhibit 1.30.

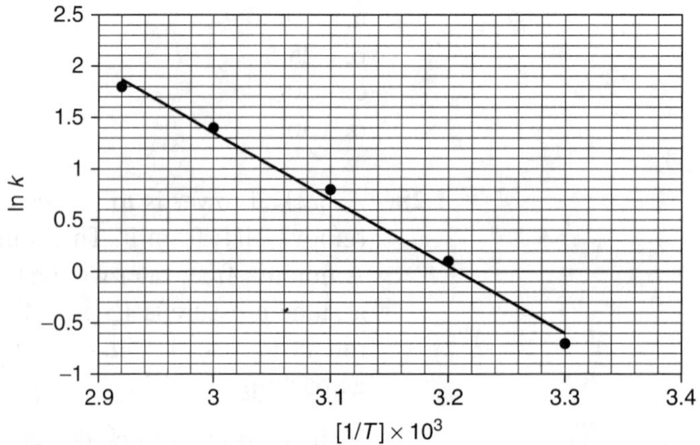

Exhibit 1.30 Graph of ln k versus [1/T] for Problem 1.30

The activation energy (cal/g mol) of this reaction is nearly:
a. 13,072
b. 20,000
c. 14,800
d. 18,600

1.31 A homogeneous liquid phase reaction

$$2A \rightarrow R \qquad -r_A = kC_A^2$$

is carried out in a mixed reactor. A 50% conversion is attained.

If the reactor is replaced by one that is six times larger, and all other factors are kept the same, the percent conversion that will be attained is nearly:

a. 67
b. 91.7
c. 95
d. 75

1.32 If the mixed reactor in Problem 1.31 is replaced by a plug flow reactor of equal size, the percent conversion attained will be nearly:

a. 67
b. 85
c. 90
d. 75

1.33 Two liquid-phase parallel reactions, of which the first one is desirable, proceed as follows:

$$A + B \xrightarrow{k_1} R + T \qquad -r_{A1} = k_1 C_A C_B^{0.2}$$

$$A + B \xrightarrow{k_2} S + U \qquad -r_{A2} = k_2 C_A^{0.5} C_B^{1.7}$$

From the point of product distribution, the most appropriate contacting scheme is:

a. C_A high, C_B low
b. C_A low, C_B low
c. C_A high, C_B high
d. C_A low, C_B high

1.34 A newly installed piece of equipment has a value of $50,000. Its useful life is estimated to be 10 years, and its salvage value is $8,000. Depreciation will be charged as a cost by making equal yearly payments, and the first payment will be made at the end of the first year. The depreciation fund will be accumulated at an annual interest rate of 8.25%. The yearly depreciation cost, in dollars, under these conditions is close to:

a. 4125
b. 2865
c. 2376
d. 3200

1.35 A corrosion allowance, typically 0.125 in. for carbon steel and 0.065 in. for stainless steel, is added to the calculated material thickness to arrive at the final plate thickness that will be used in the design of the equipment. The type of corrosion against which this allowance is used is:

a. Pitting
b. Galvanic corrosion
c. Uniform corrosion attack on the metal surface
d. Crevice corrosion

1.36 A control system has the following transfer function:

$$\frac{Y(s)}{X(s)} = \frac{\tau_1 s + 1}{\tau_2 s + 1}$$

A unit step change is applied to the system. If $\tau_1/\tau_2 = 5$, the maximum value of $y(t)$ is:

a. 0
b. 5
c. 1
d. 4

1.37 A combustibles-air mixture has the composition and the lower flammability limits (LFLs) listed in the following table:

Component	Mol %	LFL %
methane (CH_4)	2.0	5.0
ethylene (C_2H_4)	0.5	2.7
hexane (C_6H_{14})	1.0	1.1
air	96.5	

The minimum oxygen concentration (MOC) for combustion is nearly:

a. 2.34%
b. 10.0%
c. 5.67%
d. 7.92%

1.38 The following information is available on pumping a liquid through a cooler:

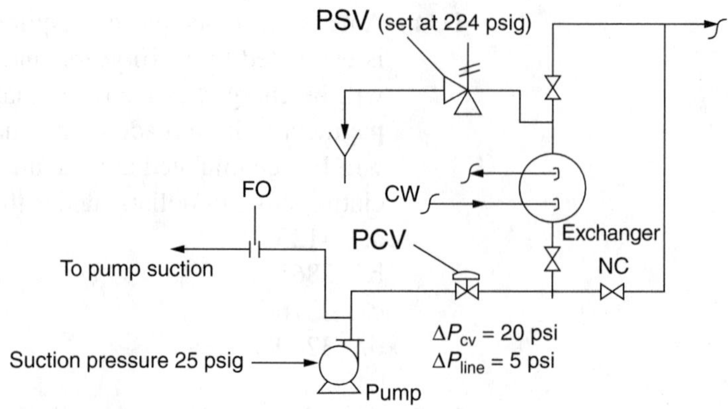

Exhibit 1.38 Process sketch for Problem 1.38

Pump normal flow = 675 gpm
Pump rated flow = 743 gpm
Pumping temperature = 330°F
Pump maximum suction pressure = 56 psig
Pump suction pressure at rated flow = 25 psig

Pump discharge pressure = 165 psig
Pump shut-off pressure = 224 psig
Exchanger shell-side inlet pressure = 140 psig
Exchanger shell-side design pressure = 224 psig = set pressure

To size the relief valve for blocked discharge, the loading (flow rate, in gpm) to be used is nearly:

a. 675
b. 743
c. 810
d. 817

1.39 The certified relieving capacity of a certain relief valve is 1000 lb/h of saturated steam. The back-pressure factor and superheat correction factors are each equal to 1. At the same pressure setting and with a back pressure that is less than 55% of the absolute relieving pressure, the capacity (lb/h) of the same valve to relieve ammonia (ratio of specific heats, $k = 1.33$) at 150°F will be nearly:

a. 1057
b. 880
c. 1136
d. 1220

1.40 A process engineer determined that the shell plate thickness should be 0.625 in. for a vessel that is 8 ft in diameter and 50 ft in tangent-to-tangent length. The design data were given as follows:

Operating pressure = 75 psig
Design pressure = 100 psig
Operating temperature = 300°F
Maximum design temperature = 650°F
Joint efficiency = 80%
Allowable stress between –20 to 650°F = 13,800 psi
Corrosion allowance used = 0.125 in.
Material of construction: SA-515-55

From a calculation of the thickness of the 2:1 ellipsoidal head for the vessel, it was found that the shell thickness controls the selection of wall thickness for the vessel. The maximum allowable working pressure (MAWP in psig) is nearly:

a. 114
b. 143
c. 133
d. 121

d. [illegible]

CHAPTER 2

Afternoon Exam

2.1 A natural gas has the following molar composition:

CH_4	97.0%
O_2	1.6%
N_2	1.4%
Total	100.0%

This gas is burned with 30% excess air in a boiler furnace. The flue gases are essentially at atmospheric pressure. If the combustion is complete, the dew point (°F) of the flue gases will be nearly:

a. 140
b. 129
c. 100
d. 120

2.2 Aluminum sulfate is manufactured by treating crushed bauxite ore with sulfuric acid. The ore contains 55.4% Al_2O_3. Sulfuric acid is 77.7% H_2SO_4. The reaction of Al_2O_3 with H_2SO_4 can be represented by the following equation:

$$Al_2O_3 + 3H_2SO_4 \rightarrow Al_2(SO_4)_3 + 3H_2O$$

The molecular weights of the compounds are as follows:

Component	Molecular wt
Al_2O_3	101.96
H_2SO_4	98.08
$Al_2(SO_4)_3$	342.14
H_2O	18.03

In one batch, 2160 lb of bauxite ore were treated with 5020 lb of sulfuric acid to obtain a final solution containing 3600 lb of aluminum sulfate. The percent conversion of Al_2O_3 is nearly:

a. 84.7
b. 67.4
c. 89.7
d. 96.2

2.3 In the Haber process to manufacture NH_3, a mixture of nitrogen and hydrogen that contains some argon as an impurity is passed through a catalyst at a pressure of 800 to 1000 atm and a temperature of 500 to 600°C. The NH_3 produced in the reactor is separated from the reaction gases in a separator. The NH_3-free gases, after a small purge to prevent the accumulation of argon, are returned to the reactor along with the feed. The product NH_3 does not contain any dissolved gases. The compositions of some of the streams are indicated in Exhibit 2.3.

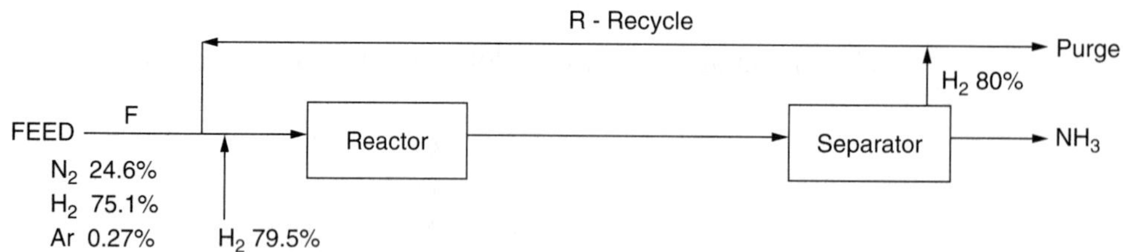

Exhibit 2.3 Process sketch of NH_3 production using Haber's process

The number of moles recycled per 100 moles of fresh feed is:

a. 932
b. 880
c. 738
d. 804

2.4 A distillation column is used to separate two components *A* (the more volatile one) and *B*. The distillate product rate is 100 kg mol/h. In the enriching (rectifying) section of the column, the liquid composition from a certain plate, *n*, (plates are counted from the top down) is found to be 0.66 mol fraction *A*, while that of the vapor from the plate immediately below it [the $(n + 1)$th plate] is 0.7514 mol fraction *A*. Assuming a constant molal overflow, the reflux ratio used in the enriching section of the column is nearly:

a. 2.5
b. 3.0
c. 3.5
d. 4.0

2.5 The Van der Waals constants for a certain gas are as follows:

$$a = 5.1 \times 10^3 \text{ psia}\left(\frac{\text{ft}^3}{\text{lb mol}}\right)^2, \qquad b = 0.516\,\frac{\text{ft}^3}{\text{lb mol}}$$

A cylinder with an internal volume of 1.82 ft^3 contains 0.0044 lb mol of this gas. A gage on the cylinder shows a pressure of 5.3 psig. The temperature (°F) of the gas in the cylinder is:

a. 102
b. 312
c. 156
d. 220

2.6 A manufacturer of high-quality lime uses precipitated, essentially pure $CaCO_3$ to produce high quality lime (CaO). In the process, $CaCO_3$, after heating to 400°F in a preheater, is fed into a rotary kiln at a rate of 20,000 lb/h. The calcined product leaves the kiln at 1832°F. Assume 100% conversion. The process is shown schematically in Exhibit 2.6.

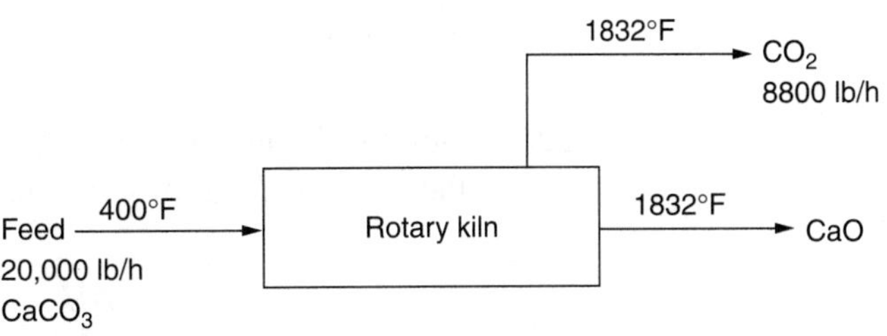

Exhibit 2.6 Process sketch for Problem 2.6

Average specific heats

Component	C_P, Btu/(lb·°F)
$CaCO_3$	0.242
CaO	0.245
CO_2	0.283

Heat of reaction: $\Delta H_r = 788.4$ Btu/lb of $CaCO_3$

If the kiln heat efficiency is 80%, then the actual heat (MMBtu/h) that needs to be supplied to the kiln is:

a. 23.4
b. 29.3
c. 27.3
d. 25.2

2.7 The combustion reactions of $CH_4(g)$ and $C_2H_6(g)$ and their heats of combustion at 25°C and 1 atm, when the products are $CO_2(g)$ and $H_2O(g)$, are given below:

$$CH_4(g) + 2O_2(g) \rightarrow CO_2(g) + 2H_2O(g)\,0 \quad \Delta H_c = -802 \text{ kJ/g mol}$$

$$C_2H_6(g) + 2O_2(g) \rightarrow 2CO_2(g) + 2H_2O(g) \quad \Delta H_c = -1428 \text{ kJ/g mol}$$

A fuel gas has the following composition:

CH_4, 85 mol%; C_2H_6, 12 mol%; N_2, 3 mol%

The higher heating value, in kJ/(100 m^3) of the fuel gas at standard conditions is nearly:

a. 44,200
b. 34,000
c. 42,100
d. 39,000

2.8 Air is preheated from 80°F to 150°F in a steam heater. The air-flow rate is 70,000 lb/h. The specific heat capacity of air can be taken to be 0.24 Btu/lb·°F. Steam enters the shell side of the heater at 300°F and leaves the heater subcooled to 280°F. The steam usage, in lb/h, for this service is nearly:

a. 1457
b. 1436
c. 1344
d. 1263

2.9 Good estimates of the heats of vaporization of liquids are obtained using the reduced form of Clausius-Clapeyron equation, which is given by

$$\frac{\Delta H_v}{ZRT_C} = \frac{B}{T_C}\left[\frac{T_r}{T_r + C/T_C}\right]^2$$

where B and C are two of the constants of the Antoine equation.

The critical constants for benzyl acetate are $P_C = 31.4$ atm, $T_C = 699$ K, and $T_B = 486$ K.

The vapor pressures of benzyl acetate can be calculated using the following Antoine equation:

$$\ln P = A - \frac{B}{T + C}$$

where T is the temperature in K, and the constants are $A = 16.5956$, $B = 4104.84$, and $C = -74.56$.

Under these conditions, the heat of vaporization (Btu/lb) of benzyl acetate at its normal boiling point is nearly:

a. 146
b. 133
c. 142
d. 123

2.10 Stoichiometric equations for some reactions are given below along with their heats of reaction (kcal/g mol). The heats of formation (kcal/g mol) of the compounds involved in the reactions are also provided.

(1) $CH_3CHO(g) + H_2(g) \rightarrow C_2H_5OH(g)$ $\qquad \Delta H_1 = -12.51$

−39.72 −52.23

(2) $C_2H_5OH(g) + 3O_2(g) \rightarrow 2CO_2(g) + 3H_2O(l)$ $\qquad \Delta H_2 = -340.83$

−52.23 −94.052 −68.314

(3) $H_2(g) + \frac{1}{2}O_2(g) \rightarrow H_2O(l)$ $\qquad \Delta H_3 = -68.32$

−68.32

(4) $H_2O(l) \rightarrow H_2O(g)$ $\qquad \Delta H_4 = +10.52$

−68.314 −57.8

From these data, the heat of combustion (kcal/g mol) of gaseous acetaldehyde at 25°C and 1 atmosphere pressure is nearly:

a. –285.02
b. –263.98
c. –318.6
d. –304.2

2.11 Soda ash liquor ($\rho = 1250$ kg/m^3 and $\mu = 1.2$ cP) is flowing through a 150-mm i.d. steel pipe at a flow rate of 125 m^3/h. Under these conditions, the Fanning friction factor is nearly:

a. 1.72×10^{-2}
b. 1.43×10^{-2}
c. 4.3×10^{-3}
d. 3.5×10^{-2}

2.12 An organic liquid ($\rho = 56.2$ lb/ft^3, $\mu = 0.65$ cP) is flowing through a standard 4-in. Schedule 40 pipe (i.d. = 0.3355 ft, relative roughness = 0.00045) at a Reynolds number equal to 2×10^5. Therefore, the pressure drop through the pipe in psi per 100 ft length of pipe is nearly:

a. 1.0
b. 1.2
c. 1.5
d. 2.0

2.13 A centrifugal pump is drawing a liquid from a horizontal tank and discharging it to another elevated point, as shown in Exhibit 2.13.

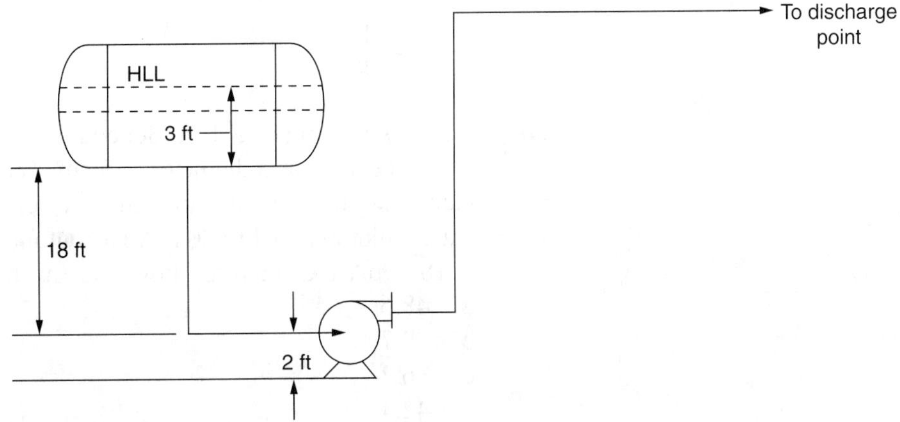

Exhibit 2.13 Sketch of pumping system for Problem 2.13

The other data for this pumping system are as follows:

Operating pressure in the tank = 26.3 psia
Design pressure of the tank = 50 psig
Specific gravity of the liquid at the pumping temperature = 0.8
Assume that the minimum suction line loss is equal to 1.5 psi.

The maximum suction pressure, in psig, at the inlet of the pump is nearly:

a. 20
b. 50
c. 70
d. 56

2.14 In Exhibit 2.14, the following curves for a pumping system are plotted: the *H-Q*, curve, the system curve, and the efficiency curve.

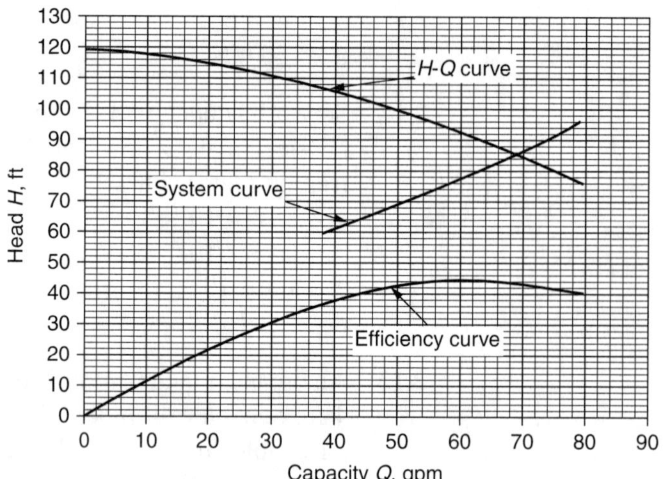

Exhibit 2.14 The *H-Q*, system, and efficiency curves for Problem 2.14

For the flow rate through the piping given by the intersection point of the system and the *H-Q* curves, the standard-size motor that should be recommended is:

a. 3 hp
b. 5 hp
c. 7.5 hp
d. 2.5 hp

2.15 Water is flowing in a 4-in. Schedule 40 pipe. The flow is measured using an orifice that has a diameter of 5 cm. The differential is measured using an electronic differential transmitter, which is calibrated for 0 to 250 cm of water column and 4 to 20 mA DC output. If the signal from the transmitter is 18.5 mA DC, then the flow rate (m^3/h) is most nearly:

a. 48.6
b. 29.7
c. 36.2
d. 42.4

2.16 The level of oil in a tank is controlled. The flow-rate range is 100 to 1200 gpm. The oil has a specific gravity of 0.9. The line pressure varies from 100 to 150 psig, and the throttling-pressure drop varies from 50 to 110 psi. The temperature varies from 70 to 150°F. The required rangeability of the control valve for this operation is:

a. 8:1
b. 10:1
c. 18:0
d. 25:1

2.17 Water is flowing through an annular channel of 50 ft length. The annular channel is made of outer pipe of rectangular tube and an inner Schedule 40, 0.5-in. pipe. The other data are as follows:

Outer pipe: 2-in. square (inside dimensions)
Inner pipe: o.d. = 0.84 in. = 0.07 ft
Flow area or channel cross section = 0.02393 ft^2
Relative roughness = 0.0015
Water flow rate = 100 gpm
Density of water = 62.4 lb/ft^3
Viscosity of water = 0.9 cP

The frictional pressure drop (psi) for 100-gpm flow in the channel is most nearly:
a. 3.2
b. 6.3
c. 12.6
d. 8.4

2.18 Lube oil is cooled in the annulus of a double-pipe exchanger from 450°F to 350°F by crude oil flowing in the tube. The following properties of lube oil are available at the caloric temperature:

Heat capacity, $C_p = 0.615$ Btu/lb·°F

Thermal conductivity, $k = 1.55 \times 10^{-6} \dfrac{\text{Btu}}{\text{s} \cdot \text{in.} \cdot °\text{F}}$

Viscosity, $\mu = 3.05$ cP

The value of the Prandtl number under these conditions is nearly:
a. 22.2
b. 57.4
c. 28.3
d. 67.7

2.19 A hollow metal (thermal conductivity k = 26 Btu/(h·ft²·°F/ft) sphere that has a 4-in. inside diameter and is 2 in. thick is heated inside so that the inside surface temperature is maintained at 300°F. If the outside surface temperature of the sphere is 220°F, the heat loss (Btu/h) from the sphere is nearly:
a. 9600
b. 8720
c. 6100
d. 7630

2.20 A bare, 2-in. steel pipe (o.d. = 2.375 in.) is carrying steam at 366°F. The temperature of the air surrounding the pipe is 70°F. The heat loss (Btu/h) per foot of pipe under these conditions will be nearly:
a. 972
b. 637
c. 856
d. 1024

2.21 Aniline is maintained at 100°F in an outdoor storage tank by passing steam through a bundle of tubes immersed in the liquid at the bottom of the tank. The horizontal cylindrical tank has a 6-in. i.d. and is 15 ft long. The tank is not insulated, but it is shielded from the wind. The radiation coefficient can be taken to be 0.75 Btu/(h·ft^2·°F). The lowest winter temperature is 0°F. Steam at 220°F is passed through the tubes (each tube o.d. = 1 in.). The properties of aniline as a function of temperature are given in the table below:

Properties of aniline

t, °F	ρ, lb/ft^3	C_p, Btu/lb·°F	μ, cP	k, Btu/(h·ft·°F)	ν, ft^3/lb
150	61.58	0.4701	1.350	0.09448	0.0162
160	61.28	0.4751	1.221	0.09379	0.0163
170	60.98	0.4797	1.115	0.0931	0.0164

The Grashof number for aniline heating is close to:

a. 6.11×10^4
b. 7.65×10^6
c. 4.96×10^6
d. 3.8×10^6

2.22 A mixture of organic vapors at a temperature of 230°F is to be condensed at 30 psia. The condensing curve was calculated, and the results are summarized in the following table.

Table 2.22 Calculations for the condensation of mixed vapors

Temperature, °F	Wt. Fraction (Vapor)	Enthalpy Change, MMBtu/h	Heat Transferred, Btu/h	Δt_W*, °F
230	1.0000	–0.0000		
220	1.0000	–0.3903	390,307	1.22
213.8	1.0000	–0.6315	241,200	0.76
210	0.8425	–2.4711	1,839,600	5.75
200	0.4813	–6.7959	4,324,800	13.53
190	0.1662	–10.6927	3,896,800	12.20
185	0.0000	–12.7833	2,090,600	6.54

*cooling water flow rate = 319,583 lb/h
Δt_w = rise in temperature of the water, °F

The cooling water is available at a temperature of 80°F. The heat load (MMBtu/h) in the condensation range is nearly:

a. 12.15
b. 12.78
c. 10.70
d. 10.06

2.23 In a one-shell-pass, one-tube-pass exchanger, the saturated steam (temperature = T_s) is condensing on the shell side and the cold fluid is being heated on the tube side from t_1 to t_2. By equating the heat picked up by the cold fluid, $WC_P(t_2 - t_1)$, and the heat transferred, $Q = UA\Delta T_{lm}$, the following relation can be obtained. (The symbols have their usual meanings.)

$$\ln\frac{T_s - t_1}{T_s - t_2} = \frac{UA}{WC_P}$$

The following operating data are provided:

$T_s = 261°F$

$t_1 = 80°F$

$t_2 = 170°F$

If the mass flow rate on the tube side is increased by 50%, while keeping all other conditions the same and assuming that the steam film and wall resistances are negligible, the new value of t_2 (the outlet temperature, °F) will be nearly:

a. 175
b. 165
c. 180
d. 155

2.24 For the diffusion of A in stagnant B, the experimental value of k_G was found to be 7.11×10^{-3} (kg mol)/(s·m^2·atm). The total pressure was 1 atm, and the temperature was 298 K.

The partial pressures at two different points were $p_{A1} = 0.2$ atm and $p_{A2} = 0.03$ atm. The value of k_G' (lb mol/h·ft^2·atm) for counter-diffusion of A and B under the same conditions of flow, temperature, and concentrations will be nearly:

a. 8.72
b. 2.40
c. 15.2
d. 4.63

2.25 The acetone-chloroform system forms an azeotrope at 64.6°C and 1 atm pressure. The Van Laar constants that are calculated using the azeotropic composition are $A = 1.7224$ and $B = -316.8$. A liquid mixture of acetone and chloroform has a boiling point of 336 K, and its composition is 57.5 mol % acetone. The vapor pressures of acetone and chloroform at 336 K are 935 and 805 mm, respectively. The relative volatility of acetone (component 1) referred to chloroform (component 2) is nearly:

a. 1.4
b. 1.2
c. 1.3
d. 1.1

2.26 A liquid mixture of carbon disulfide and carbon tetrachloride containing 67 mol % CS_2 is to be continuously separated at a pressure of 1 atmosphere. The feed rate to the column will be 100 lb mol/h. The distillate product will be 97mol % carbon disulfide. The residue will contain 1 mol % carbon disulfide. The feed to the tower will be 30 mol % vapor, and the rest will be saturated liquid. A total condenser will be used, and the reflux will be returned to the column at the bubble point. The equilibrium data for the system carbon disulfide (component A) and carbon tetrachloride (component B) is given in Exhibit 2.26.

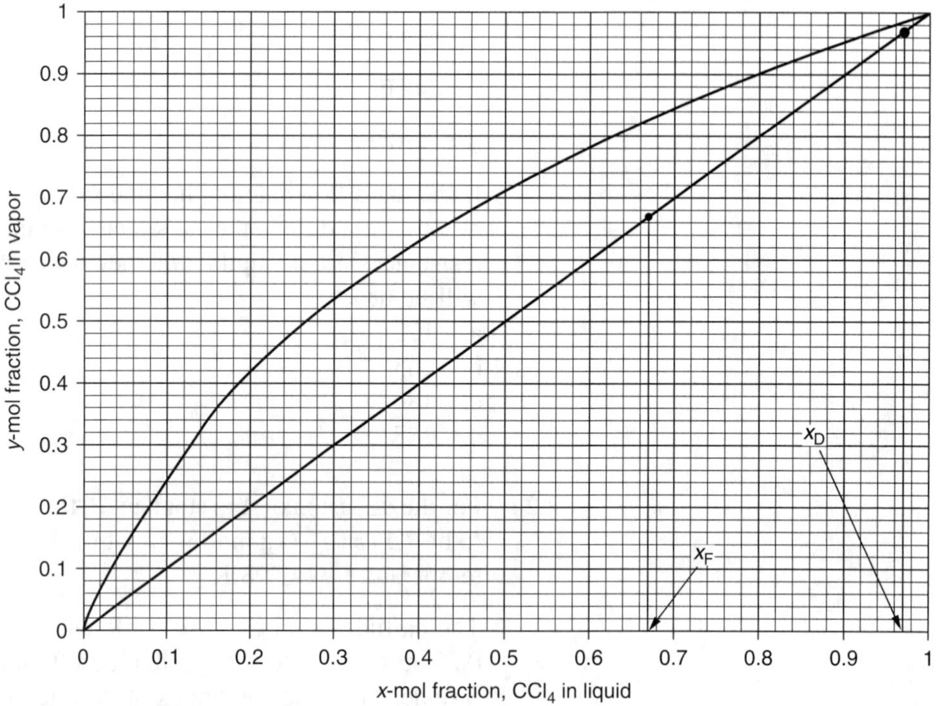

Exhibit 2.26 Equilibrium diagram for the CS_2–CCl_4 system

If twice the minimum reflux ratio is to be used for the design of the column, the actual reflux ratio will be nearly:

a. 1.846
b. 0.920
c. 2.76
d. 2.30

2.27 An existing 6-ft diameter tower will be used to absorb NH_3 from an air stream at 16 psia and 80°F that contains 10 mol % NH_3. The feed to the tower will be 1200 lb mol/h, and 99% of the NH_3 will be recovered. The ammonia-free water flow rate will be equal to 1.5 times the minimum. Also, 2-in. Pall rings ($F_p = 26$) will be used as packing.

The equilibrium relationship for the range of concentrations involved can be represented by the following equation:

$$y = 1.406x$$

where y = mol fraction of NH_3 in vapor and x = mol fraction of NH_3 in liquid.

The temperature in the tower will be maintained constant at 80°F by means of cooling coils.

The water rate (gpm/ft^2) that will be used is most nearly:
a. 5.6
b. 7.4
c. 3.2
d. 6.1

2.28 The following table contains the material balance and the molar compositions of the components for a specified separation by fractionation.

	Feed		**Distillate**		**Bottoms**	
Component	**mol**	**mol fraction**	**mol**	**mol fraction**	**mol**	**mol fraction**
butane	40	0.40	40	0.6178	0.00	0.0000
pentane	25	0.25	23.75	0.3668	1.25	0.0355
hexane	20	0.20	1.00	0.0154	19.00	0.5390
heptane	15	0.15	0.00	0.0000	15.00	0.4255
total	100	1.00	64.75	1.0000	35.25	1.0000

The column uses a total condenser, and the reboiler can be considered to be one theoretical stage. A total of 17 theoretical stages are calculated using the Erbar-Maddox correlation between the reflux ratio and the number of stages, with $R_M/(R_M + 1)$ as the parameter.

Using the Kirkbride equation, the feed will be located on which of the following theoretical stages. Assume that the plates are counted from the top to the bottom.
a. 9.0
b. 9.7
c. 8.7
d. 7.7

2.29 For the reaction $A \rightarrow B$, the following data are available:

at 25°C, Heat of reaction $\Delta H_r^0 = -23.99$ kcal/g mol

Free energy of reaction $\Delta G_f^0 = -3.0$ kcal/g mol

For the temperature range of interest, ΔH_r can be assumed to be constant.

The equilibrium constant at a temperature of 50°C is nearly:
a. 5.1
b. 6.9
c. 158.2
d. 79.3

2.30 The reaction $A \rightleftarrows B$ is first order in both directions. Its equilibrium constant is 6.9 at 50°C. If the initial concentration of B is 0, the percent equilibrium conversion at 50°C is close to:

a. 87.34
b. 79.1
c. 67.2
d. 98.3

2.31 The decomposition of benzene diazonium chloride into chlorobenzene and nitrogen is a first-order reaction. A plot of log k versus $(1/T) \times 10^3$, where k is the reaction rate constant (s^{-1}) and T is the temperature in K, is shown in Exhibit 2.31.

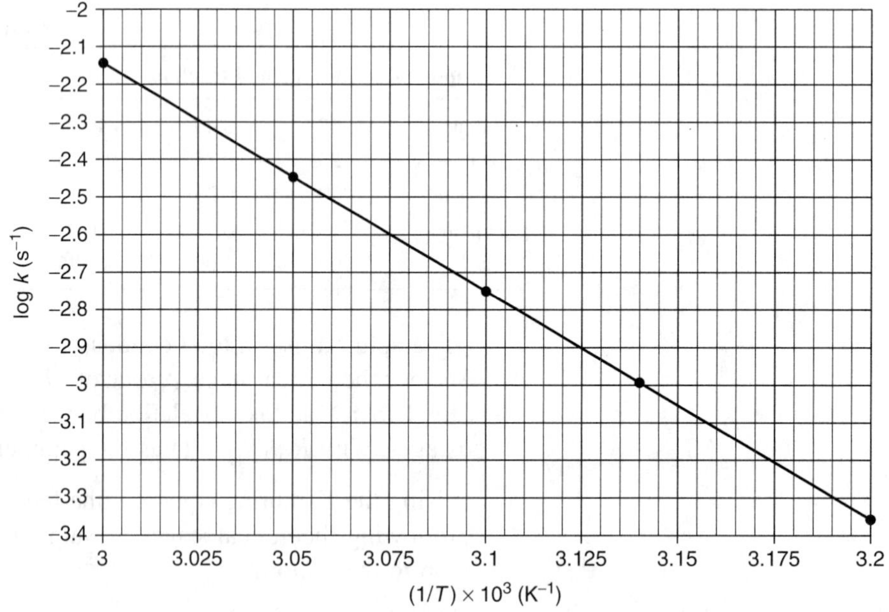

Exhibit 2.31 A plot of the logarithm of the rate constant versus the inverse of the temperature for Problem 2.31

A straight line is obtained. Calculation of the activation energy (kJ/mol) using this plot gives a value of:

a. 76.3
b. 28.7
c. 100
d. 117.5

2.32 The first-order liquid-phase reversible reaction $A \underset{k_2}{\overset{k_1}{\rightleftarrows}} R$ takes place in a batch reactor.

After 8 minutes, the conversion is 33.3%. The equilibrium conversion is 66.7%. The specific rate constant (min^{-1}) for the reverse reaction is nearly:

a. 0.1
b. 0.029
c. 0.29
d. 0.32

2.33 Hydrolysis of acetic anhydride is carried out at 25°C in four mixed-flow reactors in series. Each reactor has a volume of 230 gallons. The flow rate is 30 gpm. The inlet composition is 1.5 lb mol/gal. The reaction is first order and has a reaction constant $k = 0.158\ \text{min}^{-1}$.

The percent conversion that will be obtained is nearly:

a. 99.5
b. 97.3
c. 95.8
d. 85.7

2.34 The cost of a carbon-steel centrifugal pump (capacity 1000 gpm × 100 psi; 100-hp motor) with all of the contact parts in carbon steel is presently $20,000. If the inflation rate is 2% per year, the installed cost, in dollars, of a similar unit (capacity 1500 gpm × 90 psi; 125-hp motor) 2 years hence will be nearly:

a. 93,600
b. 102,800
c. 87,900
d. 83,300

2.35 A chemical company produces chlorinated and fluorinated chemicals. At present reactor pressure vessels are constructed of alloy C-276 (Ni = 57%, Cr = 16%, Mo = 16%, Fe = 5%). Because of excessive corrosion, they have to be replaced every 12 to 14 months. The corrosion rate of C-276 in a solution of 10% H_2SO_4 + 1% HCl at 90°C is 0.041 in. per year; the corrosion rate of alloy 59 is 0.003 in. per year in the same medium. The process involves chemicals such as hydrocarbons, ammonium fluoride, sulfuric acid, and others. A suitable new construction material that could be used to replace C-276 and prolong the reactor life under the same conditions would be:

a. Glassed-lined steel
b. Tantalum-clad vessel
c. Alloy 59 (Ni = 59%, Cr = 23%, Mo = 16%, Fe = 1%)
d. SS 317L

2.36 A first-order irreversible reaction $A \rightarrow B$ is carried out in a stirred-tank reactor. By considering the unsteady material balance over the reactor, the following relation is obtained:

$$V\frac{dC_A}{dt} + (F + kV)C_A = FC_{AO}$$

where

V = Hold-up volume of the reaction mass, ft^3
k = First-order reaction rate constant, h^{-1}
C_A = Concentration of species A in the reactor, or exit concentration, mol/ft^3
C_{AO} = Feed concentration, mol/ft^3
F = Feed rate, $\text{ft}^3\text{/h}$

In the derivation of the above equation, it is assumed that both V and ρ, the density of solution, are constant.

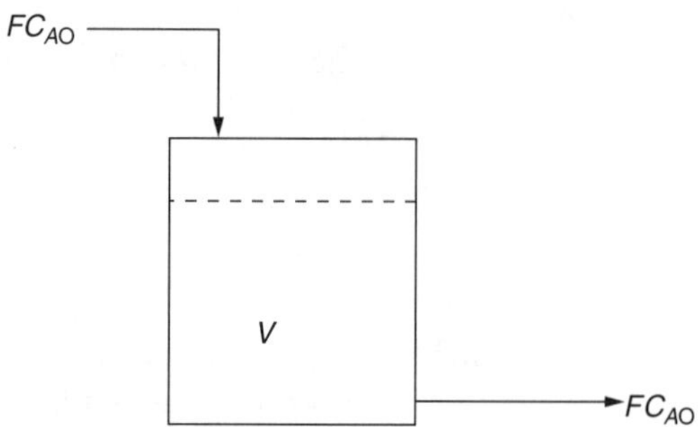

Exhibit 2.36 Process sketch for Problem 2.36

If the hold-up time is 1.6 h, and the reaction rate constant is 2 h^{-1}, the time constant τ of the reactor is nearly:

a. 0.28
b. 0.38
c. 0.34
d. 0.52

2.37 A sieve tray column is to be designed for the following conditions:

Vapor flow rate = 40,565 lb/h
Liquid flow rate = 45,620 lb/h
Density of vapor = 0.21 lb/ft^3
Density of liquid = 52.8 lb/ft^3
Surface tension of liquid = 18 dyn/cm
Tray spacing = 24 in. (allow 3 in. for liquid level and plate thickness)
Use a factor of 0.95 for foaming
Use a factor of 0.9 to account for 10% of the area for down-comers
Assume factor = 0.8 to operate the column at 80% flood

The approximate diameter of the column (ft) will be:

a. 6
b. 5.5
c. 5
d. 4.5

2.38 A vessel is to be constructed of carbon steel (SA-515-15). The following data are available:

Diameter of the tank = 6 ft
Tangent-to-tangent length = 14 ft
Operating temperature = 300°F
Design pressure = 115 psig (internal)
Design temperature = 650°F
Vessel head type – 2:1 Ellipsoidal

Joint efficiency = 0.85
Allowable stress at 650°F = 13,800 psi
Corrosion allowance = 0.125 in.
Vessel-head type is 2:1 ellipsoidal

The maximum-allowable working pressure (psig) for this vessel is nearly:
a. 127
b. 139
c. 149
d. 115

2.39 A steel vessel containing a flammable liquid is installed as shown in Exhibit 2.39.

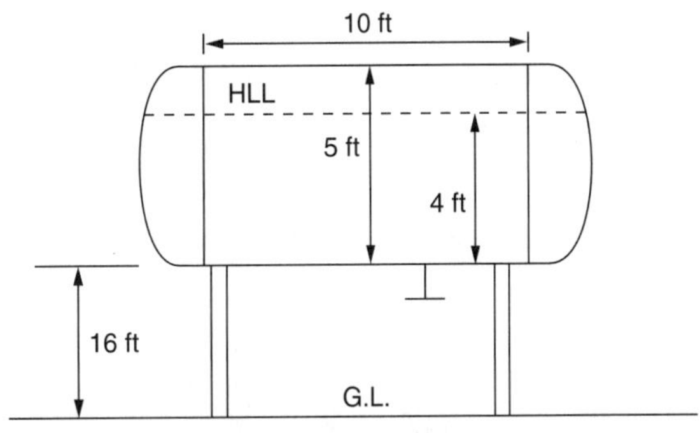

Exhibit 2.39 Layout of storage vessel for Problem 2.39

The vessel is uninsulated, and there is no fire-fighting provision such as water spray. The fire heat load (MMBtu/h), in accordance with API 2000, will be nearly:
a. 4.7
b. 4.18
c. 1.7
d. 4.55

2.40 The 5-day BOD of an industrial waste is 250 mg/L. The first-stage ultimate O_2 demand is 350 mg/L at 20°C. The time required to satisfy the 50% O_2 demand is nearly:
a. 4.8 days
b. 6.4 days
c. 3.2 days
d. 2.8 days

CHAPTER 3

Additional Review Problems

OUTLINE

MASS/ENERGY BALANCES AND THERMODYNAMICS

Problems 3.1–3.2

Dichloroacetyl chloride is produced by reacting trichloroethylene with oxygen according to the following reactions (TCE—trichloroethylene; DCAC—dichloroacetyl chloride):

Step 1. Slow reaction:

$$\underset{\text{TCE}}{2CHCl{=}CCl_2} + O_2 \xrightarrow[\substack{Cl_2 \\ 60-85^\circ C}]{UV} \underset{\text{DCAC}}{CHCl_2COCl} + \underset{\text{Epoxide}}{CHClOCCl_2}$$

Step 2. Fast reaction:

$$\underset{\text{Epoxide}}{CHClOCCl_2} \xrightarrow[60-85^\circ C]{DMF} \underset{\text{DCAC}}{CHCl_2COCl}$$

Molecular weights of the reactants and products are given below.

Component	Mol wt.
TCE	131.5
DCAC	147.5
O_2	32

Yield based on TCE = 84.9%.
Yield based on oxygen = 60.3%.

3.1 TCE (lb) needed per 100 lb of DCAC is most nearly:
a. 890
b. 1100
c. 950
d. 1050

3.2 Oxygen used (lb) per 1000 lb of DCAC is most nearly:
a. 108
b. 180
c. 165
d. 150

3.3 A gas is composed of 90% CH_4 and 10% N_2 by volume. It is burned under a boiler. The Orsat analysis of the flue gas leaving the boiler shows CO_2, 9.08%; O_2, 4.54%; and N_2, 86.38%.

Assuming complete combustion in the furnace, the percent excess air used is most nearly:
a. 20
b. 22.5
c. 25
d. 28.6

3.4 In a Haber process to manufacture NH_3 a mixture of nitrogen and hydrogen, which contains some argon as impurity, is passed through a catalyst at 800 to 1000 atm and at a temperature of 500 to 600°C. The NH_3 produced in the reactor is separated from the reaction gases in a separator. NH_3 free gases, after a small purge to prevent accumulation of argon, are returned to the reactor along with the feed. The NH_3 product does not contain any dissolved gases. The compositions of some streams are indicated in Exhibit 3.4.

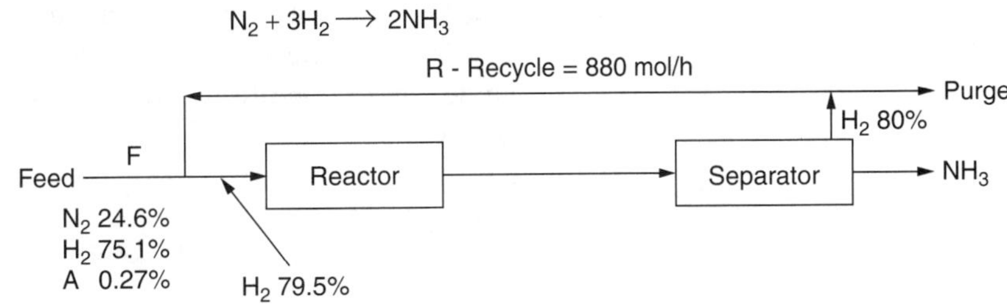

Exhibit 3.4 Process sketch for Problem 3.4

The percent conversion of hydrogen per pass is most nearly:

a. 12.23
b. 9.2
c. 11.4
d. 10.8

3.5 A mixture of N_2 and n-hexane is fed to a condenser at a flow rate of 10,000 kg/h. The composition of the gas is 20 mol % hexane, and the rest nitrogen. The gaseous effluent from the condenser is 5% hexane, as shown in Exhibit 3.5.

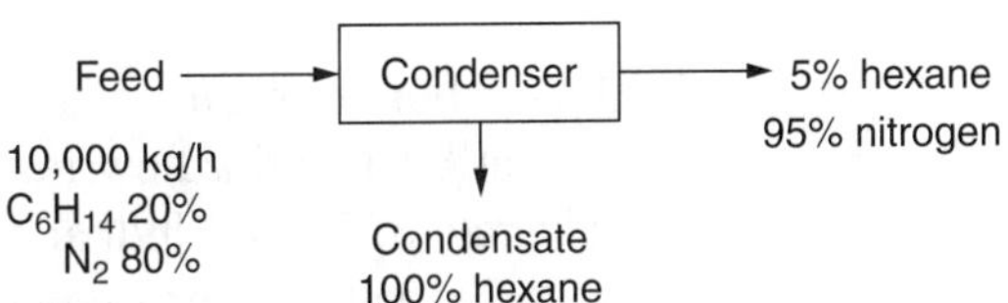

Exhibit 3.5 Process sketch for Problem 3.5

The amount of hexane (lb/h) recovered as condensate is most nearly:

a. 4200
b. 2035
c. 2546
d. 3430

3.6 Upon cooling of 1500 ft^3 of air saturated with water vapor at 30°C and 740 mm Hg to a lower temperature, 75% of the water condenses out. The amount of water condensed (lb) is most nearly:

a. 2.13
b. 3.4
c. 1.7
d. 2.57

3.7 A 1000-ft^3 tank contains 55 lb of CO_2, 140 lb of N_2, and 32 lb of O_2. The temperature of the gas is 120°F. Assume ideal gas behavior. Under these conditions, the gage attached to the vessel will show a pressure nearest to:

a. 45.1
b. 30.4
c. 40.3
d. 35.7

3.8 A 2-m^3 tank contains ammonia at 60°C and a pressure of 24 atm. The critical constants of ammonia are P_C = 111.5 atm, and T_C = 132.4°C. Under these conditions, the mass of ammonia (kg) in the tank is most nearly:

a. 29.7
b. 32.1
c. 35.5
d. 39.3

3.9 An ideal gas at a temperature of 25°C is contained in a cylinder fitted with a movable piston. The cylinder is placed in boiling water with the piston held in position until its contents rise to 100°C. During this process 2 kcal of heat is transferred to the gas. After the contents reach 100°C the piston is released and the gas does work equal to 100 J. The final gas temperature is 100°C. The total heat (J) transferred to the gas during the two steps is most nearly:

a. 8370
b. 100
c. 8470
d. 8270

3.10 A boiler produces saturated steam at 230.3 psig. Two water streams—one at 80°F and a flow rate of 16,500 lb/h and the other at 150°F and a flow rate of 23,500 lb/h are fed to the boiler. The boiler pressure is 230.3 psig. The velocity of steam in the boiler exit pipe is 630 ft/s.

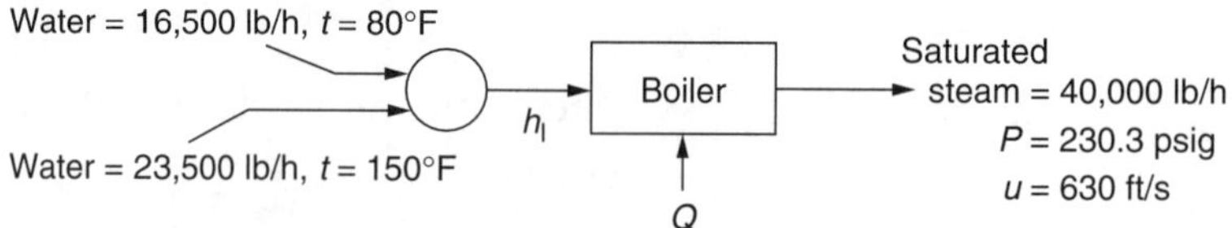

Exhibit 3.10 Boiler sketch for Problem 3.10

The heat (MMBtu/h) to be supplied to the boiler is most nearly:

a. 40.1
b. 45.3
c. 50.6
d. 55.9

3.11 The following data for 1,3-butadiene are reproduced from Perry's handbook:

Temperature, °F	Pressure, psia	Volume, ft^3/lb		Enthalpy, Btu/lb	
		Liquid	Vapor	Liquid	Vapor, Saturated
–40	2.867	0.02320	28.75	180.29	372.7
0	8.461	0.02406	10.525	199.94	383.9
50	24.94	0.02529	3.840	225.66	398.2
120	80.11	0.02427	1.262	264.60	418.2

If 1,3-butadiene liquid in a closed vessel and initially at –40°F is allowed to reach a saturated vapor state at 120°F, the change in internal energy (Btu/lb) of 1,3–butadiene will be nearest to:

a. 237.91
b. 219.06
c. 180.31
d. 190.7

3.12 A distillation column is fractionating a benzene-toluene mixture. The distillate product composition is 97 mol % benzene and 3 mol % toluene. A reflux ratio of 2 is used. Reflux is at its bubble point. Latent heat of vaporization of benzene is 30.76 kJ/g mol and that of toluene is 33.5 kJ/g mol.

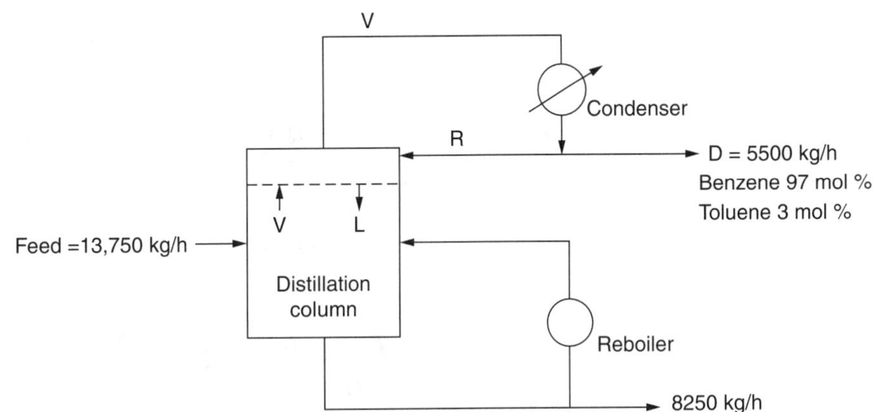

Exhibit 3.12 Sketch of distillation column for Problem 3.12

The condenser duty (kJ/h) is nearly:

a. 7.10×10^6
b. 6.1×10^6
c. 6.7×10^6
d. 6.49×10^6

3.13 When the heat of vaporization is not readily available, one can use Chen's equation to calculate the heat of vaporization within 2% of its actual value at the boiling point. The heat of vaporization at other temperatures can be then calculated using Watson's equation. The following data for ethyl bromide are available in the literature: BP = 38.2°C, T_C = 504 K, and P_C = 61.5 atm.

The heat of vaporization of ethyl bromide at 0°C (kJ/kg mol) calculated by using Chen's equation to calculate the heat of vaporization at the boiling point, followed by the use of Watson's equation, is near to:

a. 30.1×10^3
b. 28.1×10^3
c. 12.71×10^3
d. 20.86×10^3

3.14 10 lb of water at 70°F in a container are heated to 1500°F and 260 psia. The enthalpy change (Btu) in this process is most nearly:

a. 1567.7
b. 17,620
c. 12,762
d. 15,678

Problems 3.15–3.16

A steam turbine operates adiabatically under the conditions shown in Exhibit 3.15. The steam flow rate is 20,000 lb/h. The steam exiting the turbine is at 20 psia, and saturated.

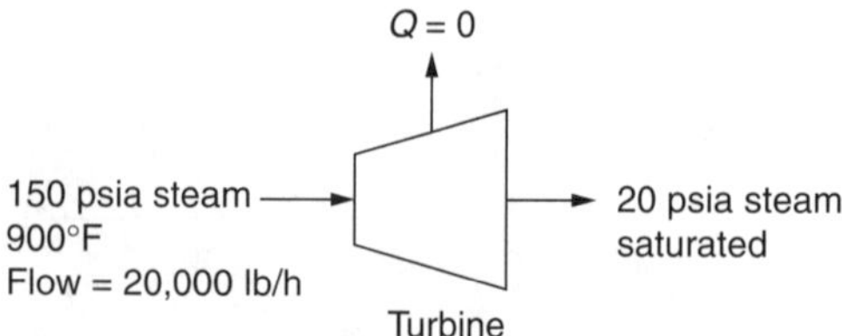

Exhibit 3.15 Process sketch for Problems 3.15 and 3.16

3.15 If the turbine efficiency is 85%, the work output (hp) of the turbine is near to:

a. 1600
b. 2200
c. 2000
d. 1800

3.16 If the steam is throttled to 100 psia and the turbine efficiency declines to 80%, the steam exit temperature (°F) would be near to:

a. 500
b. 600
c. 565
d. 635

Problems 3.17–3.18

Ammonia is to be compressed from 14 psia and 50°F to a pressure of 60 psia in a single stage. The properties of ammonia at the inlet of the compressor are given in the diagram below.

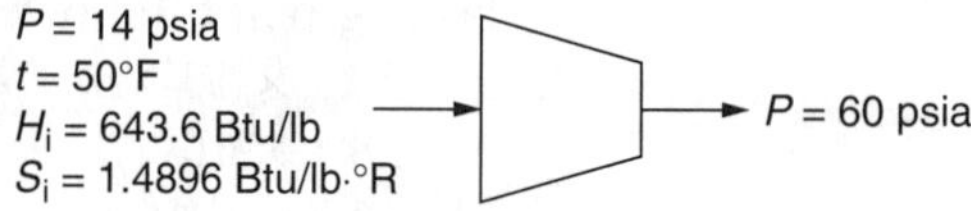

Exhibit 3.17 Process sketch for Problems 3.17 and 3.18

The properties of ammonia at 60 psia can be calculated from the following table.

	60 psia		
Temp, °F	v, ft^3/lb	H, Btu/lb	S, Btu/lb·°F
220	7.019	728.6	1.4658
240	7.238	739.7	1.4819
260	7.457	750.9	1.4976
280	7.675	762.1	1.5130
300	7.892	773.3	1.5281

3.17 If the compressor efficiency is 80%, the actual work (Btu/lb of ammonia) done by the compressor is most nearly:

a. −105.8
b. +105.8
c. −102
d. −132

3.18 Entropy increase (Btu/lb·°R) in the compression process is near to:

a. 0.4896
b. 0.0349
c. 0.00315
d. 0.217

3.19 The dehydrogenation of ethyl alcohol proceeds according to the following reaction:

$$C_2H_5OH(v) \rightarrow CH_3CHO(v) + H_2(g)$$

The heats of formation of the reactants and products are

$$C_2H_5OH(v) \quad \Delta H_f = -235.31 \text{ kJ/g mol}$$
$$CH_3CHO(v) \quad \Delta H_f = -166.2 \text{ kJ/g mol}$$
$$H_2(g) \quad \Delta H_f = 0 \text{ kJ/g mol}$$

The heat of reaction of ethyl alcohol dehydrogenation **at constant volume**, in units of kJ/mol, is most nearly:

a. 69.11
b. −66.63
c. 66.63
d. −69.11

3.20 A fuel has the composition of 80% CH_4, 15% C_2H_6, and 5% C_3H_8. Some additional data are given below.

Component	Mol wt.	Heat of Combustion, Btu/lb mol
CH_4	16	−345,776
C_2H_6	30	−615,672
C_3H_8	44	−887,776

The higher heating value of this fuel (Btu/lb fuel) is near to:
a. 19,275
b. −19,275
c. 23,362
d. −23,362

3.21 The following data are available with respect to benzyl acetate.

$P_C = 31.4$ atm, $T_C = 699$ K, normal BP = 213.5°C

Heat of vaporization at normal BP = 11,035 cal/g mol

The entropy change, in Btu/lb·°R, when benzyl acetate is vaporized to saturated vapor from its saturated liquid state at 25°C is near to:
a. 0.157
b. 0.314
c. 0.183
d. 0.266

3.22 Watson's correlation calculates the heat of vaporization of a liquid at another temperature if the heat of vaporization is available at one temperature. Heat of vaporization of propyl alcohol at its boiling point (t_b = 97.2°C) is 164.36 cal/g. Its critical temperature is 263.7°C. Its heat of vaporization (J/g mol) at 115°C is most nearly:
a. 27,460
b. 36,840
c. 47,573
d. 39,625

3.23 Specific volume of superheated SO_2 at 1000 psia and 480°F is 0.1296 ft^3/lb. Its residual volume (ft^3/lb) at these conditions is most nearly:
a. 0.041
b. 0.028
c. 0.062
d. 0.057

3.24 A vapor compression system uses HFC-134a as a refrigerant. The condenser operates at 150 psia, and the refrigerant leaves the condenser as saturated liquid. The evaporator temperature is 0°F. The refrigeration cycle uses an adiabatic turbine with 85% efficiency. The entropy increase [Btu/(lb·°R)] during turbine expansion is:
a. 0.013
b. 0.006
c. 0.00
d. 0.0012

Hint: Use tables for HFC-134a.

3.25 The entropy change, [Btu/(lb mol·°R)] that takes place when one lb mol of ethyl ether originally at 68°F and 1 atmosphere is heated to a state of superheated vapor at 122°F and 1 atm pressure is near to:

a. 5.8
b. 0.32
c. 12.40
d. 23.67

The properties of ethyl ether are as follows:

Specific heat of liquid = 0.521 Btu/(lb·°F)
Boiling point at 1 atm = 94.3°F
Molecular weight = 74.12
Heat of vaporization = 151.06 Btu/lb
Specific heat of vapor = 0.44 Btu/(lb·°F)

(Assume specific heats to be constant.)

3.26 A rigid vessel contains 1 lb of a mixture of water and steam at 150 psia. Heat is added to the vessel until the contents of the vessel reach the condition of 500 psia and 700°F. The properties of water and steam extracted from steam tables for the two conditions are given below.

Initial Condition	**Final Condition**
150 psia	500 psia
$t = 358.42°F$	$t = 700°F$
$\hat{V}_{\ell l} = 0.01809$ ft³/lb	$\hat{V}_{\ell 2} = 0$ ft³/lb. no liquid present
$\hat{V}_{gl} = 3.015$ ft³/lb	$\hat{V}_{g2} = 1.3044$ ft³/lb
$\hat{H}_v = 1194.1$ Btu/lb	$\hat{H}_v = 1357.0$ Btu/lb
$\hat{H}_l = 330.51$ Btu/lb	No liquid present; superheated vapor

The heat (Btu/lb) required to be added to the vessel to reach the final state is near to:

a. 602.4
b. 632.6
c. 571.3
d. 655.8

3.27 A Carnot heat pump is used for heating a building. The outside air is at 30°F and is the cold reservoir. The building is to be maintained at 75°F. 200,000 Btu/h are required for the heating. The heat (Btu/h) taken from outside is near to:

a. 183,173
b. 200,000
c. 100,000
d. 167,000

3.28 Ethylene at 500 atm and a temperature of 100°C is contained in a cylinder of internal volume 2 ft^3. The weight (lb) of ethylene contained in the cylinder is near to:

a. 65.4
b. 60.4
c. 68.6
d. 62.8

3.29 A Carnot engine that receives heat at a certain temperature develops 2 hp and rejects 7500 Btu/h to a sink at 60°F. The heat source temperature (°F) of the Carnot engine is near to:

a. 400
b. 450
c. 413
d. 520

3.30 One mol of an ideal gas is undergoing a reversible isothermal process. The value of $W = \int_1^2 PdV$ in this process is most nearly:

a. 0
b. $\frac{P_2V_2 - P_1V_1}{k-1}$
c. $P_1V_1 \ln\frac{V_2}{V_1}$
d. $P_1(V_2 - V_1)$

FLUID MECHANICS

3.31 For measuring liquid level in an inaccessible underground tank, the scheme shown in Exhibit 3.31 is employed. The manometer fluid has a density of 2.5 g/mL while that of the liquid in the tank is 1.05 g/mL. Air flow is controlled so that the frictional, contraction, and expansion losses in the feed line amount to 1.34" head of liquid in the tank. The velocity in the 1/4" tubing is 2.9 ft/s and that in the 1/2" pipe is 0.28 ft/s. Density of air = 0.0754 lb/ft^3.

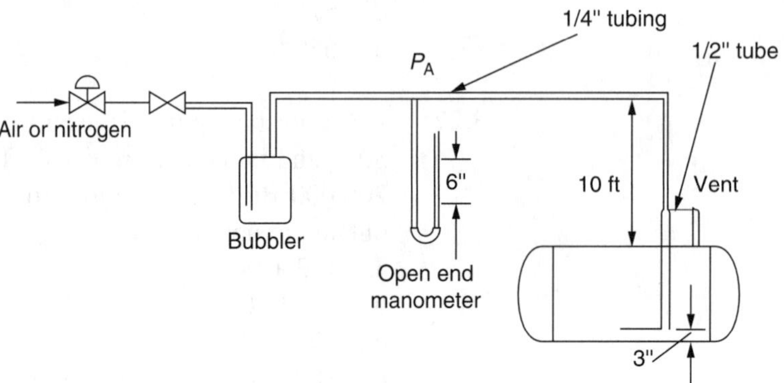

Exhibit 3.31 Sketch for Problem 3.31

If the manometer shows a 6" differential, the level (in.) of liquid from the bottom of the tank is near to:

a. 20
b. 16
c. 13
d. 17

3.32 Oil is flowing in a 2-inch-nominal-diameter, 400-feet-long pipe. The data on the pipe are as follows:

Pipe ID = 2.067" = 0.17225 ft.

Inside area of cross section = 0.0233 ft^2

Relative roughness of pipe = 0.0009

Density of oil = 51 lb/ft^3, Viscosity of oil = 2.2 cP.

If the pressure drop in this line is not to exceed 7.5 psi, the allowable velocity (ft/s) through the pipe is near to:

a. 6.0
b. 3.0
c. 4.7
d. 8.0

3.33 Air is flowing in a 20" ID duct at 220°F and 750 mm pressure. A pitot tube positioned at the center of the pipe's cross section shows a manometer reading of 4" water column. The pitot tube coefficient is 0.98. Viscosity of air = 0.022 cP. The air flow rate, expressed in scfm at 60°F and 1 atm through the duct is most nearly:

a. 16,022
b. 12,093
c. 15,000
d. 10,500

Problems 3.34–3.37

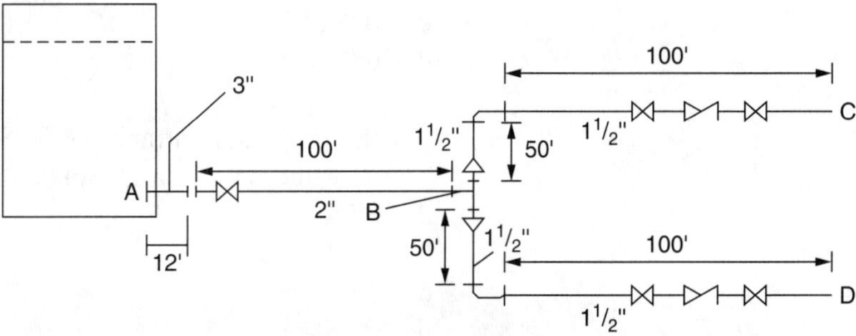

Exhibit 3.34 Water discharge from a tank to two points

As shown in the Exhibit 3.34, water at 70°F is discharged from a tank to two points, C and D. Flow through each $1^1/_2$" line is 20 gpm. Points A, B, C, and D are at the same elevation. Additional data are as follows:

Density of water = 62.3 lb/ft^3
Viscosity of water = 0.982 cP

Dia. (nom)	ID in.	ID ft	A_C ft^2	u ft/s	$\in/d_i$	Re No.	f'
3"	3.068	0.2557	0.05134	1.74	0.00059	4.2×10^4	0.024
2"	2.067	0.1753	0.02330	3.83	0.00087	6.23×10^4	0.023
$1^1/_2$"	1.61	0.1342	0.01415	3.15	0.00112	3.99×10^4	0.024

The following nomenclature is defined for the section AB of the piping:

ΣK = Sum of the resistance coefficients representing head loss due to fittings, contraction, and expansion
L = Length of piping in section AB, based on 2" Schedule 40 pipe
d_i = inside diameter, in.
f' = Moody friction factor
f = Fanning friction factor
u = velocity, ft/s
h_f = head loss due to friction, ft of liquid

3.34 The correct relationship to be used to calculate h_f for section AB of the piping is:

a. $h_f = \dfrac{2fL}{d_i}\dfrac{u^2}{g_c}$

b. $h_f = \left(\Sigma K + \dfrac{4f'L}{d_i}\right)\dfrac{u^2}{2g_c}$

c. $h_f = \left(\Sigma K + \dfrac{4fL}{d_i}\right)\dfrac{u^2}{2g_c}$

d. $h_f = \dfrac{f'L}{d_i}\dfrac{u^2}{2g_c}$

3.35 Fanning friction factor for the flow in 2" pipe is most nearly:
a. 0.00023
b. 0.00575
c. 0.025
d. 0.023

3.36 For the projected entrance of 3" Schedule 40 pipe in the tank, $K = 0.78$. The K value based on 2" Schedule 40 pipe will be most nearly:
a. 0.78
b. 0.53
c. 1.0
d. 0.16

3.37 The Reynolds number for the $1^1/_2$" Schedule 40 pipe is most nearly:
a. 1.1×10^5
b. 8.0×10^4
c. 3.99×10^4
d. 2×10^4

3.38 Water is flowing in a 3" Schedule 40 pipe at a rate of 34.1 m^3/h. The specific gravity of water = 1 and its viscosity = 1 cP. A $1^3/_4$" diameter orifice is mounted in the pipe to measure the water flow. A mercury manometer connected across the orifice will show a differential height (cm Hg) most nearly:

a. 40.5
b. 38.1
c. 36.2
d. 30.6

Problems 3.39–3.42

A liquid is pumped from a horizontal tank and delivered to a column elsewhere as shown in Exhibit 3.39.

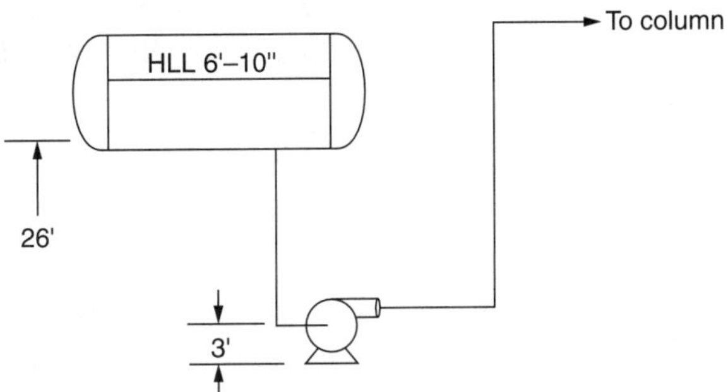

Exhibit 3.39 Schematics of pumping system, Problems 3.39–3.42

The other data are as follows:

Operating pressure = 0 psig
Design pressure = 50 psig.
Operating temperature = 181°F
Vapor pressure at normal pumping temperature = 14.7 psia
Specific gravity of liquid = 0.81
Discharge pressure = 190 psig

3.39 The total dynamic head of the pump (ft) is most nearly:

a. 400
b. 450
c. 479
d. 523

3.40 Available NPSH (ft) is near to:

a. 15
b. 16.4
c. 18.8
d. 23.1

3.41 The maximum suction pressure (psig) is near to:

a. 61.5
b. 59
c. 61
d. 57.6

3.42 A centrifugal pump takes brine from a tank and delivers it to another tank situated at an elevation of 250 ft. The pump discharge line is 8-in. nominal diameter and 400 ft long. The sum of the resistance coefficients of all fittings in the line and a sudden expansion is $\Sigma K = 8.22$. The flow rate is 1350 gpm. The density of the brine is 74.9 lb/ft^3, and the viscosity is 1.2 cP. The ID of pipe = 0.6651 ft, and the flow area = 0.3474 ft^2.

The total equivalent length of the pump discharge line (ft) is most nearly:

a. 600
b. 650
c. 760
d. 826

3.43 Three tanks, A, B, and C, at different elevations are interconnected as shown in Exhibit 3.43. The levels in the tanks are shown, as is the level in the piezometric tube. Flow rate quantities are also indicated.

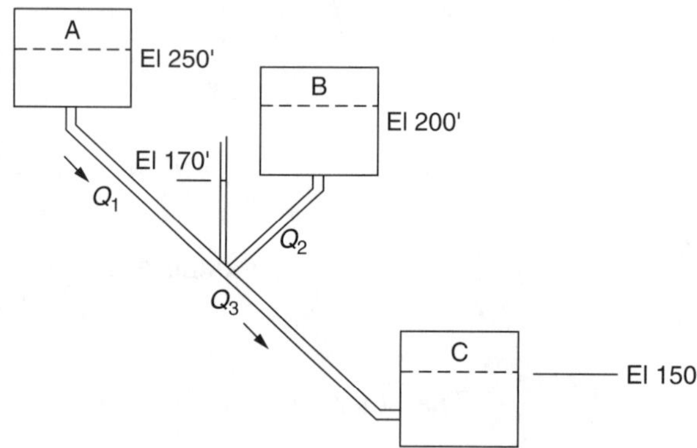

Exhibit 3.43 Sketch for Problem 3.43

Under the above conditions, the flows Q_1, Q_2, and Q_3 are related by:

a. $Q_1 + Q_2 = Q_3$
b. $Q_1 - Q_2 = Q_3$
c. $Q_2 - Q_1 = Q_3$
d. $Q_1 + Q_2 + Q_3 = 0$

3.44 Air at 80°F and 0 psig is to be compressed to 35 psig adiabatically at the rate of 16.67 lb/min. The specific heat ratio, k, for air can be taken as 1.4. Assuming $z = 1$, the adiabatic horsepower for this compression is near to:

a. 35
b. 21
c. 26
d. 31

3.45 For the adiabatic compression in Problem 3.44, the theoretical outlet temperature (°F) of air is near to:

a. 304.8
b. 764.8
c. 425
d. 471

3.46 In Problem 3.44, if the adiabatic compression efficiency is 80%, the actual discharge temperature (°F) will be most nearly:

a. 821
b. 361
c. 456.1
d. 548

3.47 In Problem 3.44, if the compression is isothermal, the theoretical horsepower (hp) required is most nearly:

a. 21
b. 24.6
c. 23.1
d. 17.6

3.48 Lube oil of density 56.1 lb/ft^3 and viscosity 470 cP is flowing through a 4" diameter, Schedule 40 pipe at a rate of 325 gpm. The linear velocity (ft/s) of oil at a point midway between the center of pipe and the wall is most nearly:

a. 8.22
b. 9.03
c. 10.7
d. 12.15

3.49 A cast iron pipe has an inside diameter of 24 inches and is running 75% full of water at a steady uniform flow as shown in Exhibit 3.49. The pipe has a slope of 3/4 in. per foot.

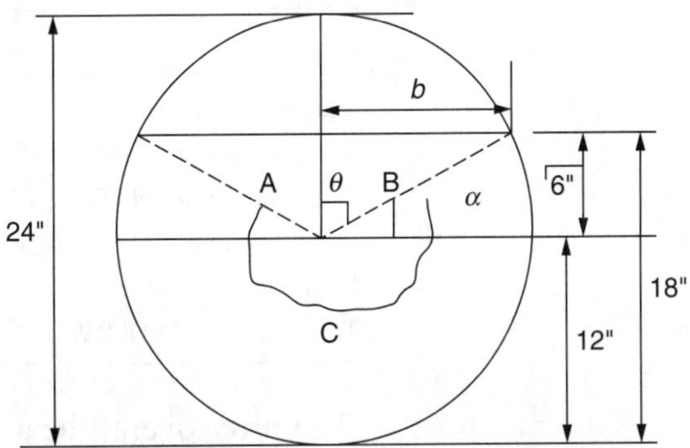

Exhibit 3.49 Sketch for Problem 3.49

The equivalent diameter (in.) to be used in fluid flow calculations of pressure drop is most nearly:

a. 24
b. 18
c. 27
d. 29

3.50 A liquid is flowing in a pipe at a certain normal flow rate. The flow is fully turbulent. If the flow rate is increased by 10% with all other conditions remaining the same, the increase in pressure drop will be near to:

a. 5%
b. 10%
c. 15%
d. 21%

HEAT TRANSFER

Problems 3.51–3.52

The composite wall of a furnace is constructed of 6" thick fire brick and common brick of certain thickness. Thermal conductivities of the firebrick and common brick are 0.08 and 0.8 Btu/(h·ft^2·°F/ft), respectively. The temperature of the inner surface of the wall is 1500°F and the outer surface of the common brick is 180°F. The heat loss from the furnace is found to be 186.4 Btu/h · ft^2.

3.51 The thickness (in.) of the common brick wall is most nearly:
a. 6
b. 9
c. 8
d. 4.5

3.52 The temperature (°F) at the interface of the firebrick and the common brick is most nearly:
a. 840
b. 335
c. 515
d. 680

3.53 A double-pipe, countercurrent heat exchanger (outside surface area = 200 ft^2) has reached its design capacity of 490,000 Btu/h and is to be shut down for cleaning. The data for the exchanger are as follows:

	Shell Side	Tube Side
Temperature of hot fluid, °F, in	350	
Temperature of hot fluid, °F, out	450	
Temperature of cold fluid, °F, in		300
Temperature of cold fluid, °F, out		310
Heat transfer coefficient, Btu/h·ft^2·°F	38.4	300
Tubes are 1" OD 18 BWG.		

The dirty overall heat transfer coefficient based on outside area is 28.03 Btu/h·ft^2·°F. If the metal wall resistance is negligible and the tube inside dirt factor is 0.003 h·ft^2·°F/Btu, the dirt factor (h·ft^2·°F/Btu) on the shell side is most nearly:
a. 0.0038
b. 0.00268
c. 0.00205
d. 0.00285

3.54 Aniline is maintained at 100°F in an outdoor storage tank by passing steam through a bundle of tubes immersed in the liquid at the bottom of the tank. The horizontal cylindrical tank is 6′ ID by 15′ long. The tank is not insulated but shielded from wind. The lowest winter temperature is 0°F. Steam at 220°F is passed through the tubes (tube OD = 1"). The properties of aniline as a function of temperature are given in the table below.

Properties of aniline

t, °F	ρ, lb/ft^3	C_p Btu/lb·°F	μ, cP	k, Btu/h·ft·°F
150	61.58	0.4701	1.350	0.09448
160	61.28	0.4751	1.221	0.09379
170	60.98	0.4797	1.115	0.0931

Hints: Kern and other authors have used the following equation to calculate free convection heat transfer coefficient on the outside of a bank of tubes:

$$h_c = 116\left[\left(\frac{k_f^3 \rho_f^2 C_f \beta}{\mu_f'}\right)\left(\frac{\Delta t}{d_o}\right)\right]^{0.25}$$

where

h_c = Convection coefficient of heat transfer, Btu/h·ft^2·°F
k_f = Thermal conductivity at film temperature, Btu/h·ft^2·°F/ft
ρ_f = density of the fluid, lb/ft^3
C_f = specific heat, Btu/lb·°F
β = Coefficient of expansion, 1/°F
μ'_f = viscosity, cP
Δt = temperature difference, °F
d_o = Outside diameter of tube, in.

Using Kern's equation for a bank of tubes, the convective coefficient by natural convection [Btu/(h·ft^2·°F)] is most nearly:

a. 60
b. 100
c. 40
d. 80

3.55 A 2" IPS steel pipe carrying steam at 300°F is insulated with $^1/_2$" of rock wool (k = 0.033 Btu/h·ft^2·°F/ft). The temperature of the wool surface exposed to surrounding air (t = 70°F) is 125°F. The emissivity of the insulation is 0.9. The radiation coefficient of heat transfer h_r [Btu/(h·ft^2·°F)] is most nearly:

a. 2.5
b. 1.1
c. 1.5
d. 1.7

3.56 Oil is flowing at a velocity of 4.5 ft/s through a 10-ft-long, 1" OD, 18 BWG, (ID = 0.902") tube. Steam is condensing on the outside surface of the tube at a temperature of 220°F. Oil enters the tube at 80°F and leaves at 100°F. Properties of oil assumed constant are

Density = 56 lb/ft^3, specific heat = 0.48 Btu/lb·°F,
Thermal conductivity = 0.08 Btu/h·ft^2·°F/ft

The viscosity of the oil varies with temperature as follows:

t, °F	80	90	100	110	120	140	220
μ, cP	20	18	16.2	15	13.5	11	3.6

The flow is expected to be streamline. The inside film coefficient of heat transfer (Btu/h·ft²·°F) is most nearly:

a. 32.2
b. 30
c. 34.1
d. 36.8

Problems 3.57–3.59

100,000 lb/h of a caustic solution (ρ = 69.6 lb·ft³) are to be cooled from 190 to 120°F. Cooling water at a rate of 154,000 lb/h enters the tubes of a 1-4 exchanger at 80°F and leaves at 120°F. The dimensions of the exchanger are as follows:

Shell Side	**Tube Side**
Shell ID = 21.25"	N_t = 172, 16′0" long
Baffle spacing = 6"	Tube OD = 1", 14 BWG, ID = 0.834"
	$1^1/_4$", triangular pitch
1 pass	4 passes

The properties of the fluids are as follows:

Caustic solution:
Specific heat = 0.88 Btu/lb·°F
Thermal conductivity = 0.34 Btu/h·ft²·°F/ft
Viscosities at different temperatures:

***t*, °F**	80	90	100	120	140	160	180	200	210	220
***μ*, cP**			1.4						0.43	

Water:

t	°F	80	90	100	120	140	160	180
μ	lb/h·ft	2.08	1.85	1.66	1.36	1.14	0.970	0.840
k	Btu/h·ft²·°F/ft	0.351	0.357	0.363	0.372	0.379	0.385	0.390

3.57 The tube heat transfer coefficient [Btu/(h·ft²·°F)] based on the inside surface is most nearly:

a. 900
b. 1050
c. 1500
d. 850

Hint: Use the Dittus-Boelter equation to calculate the heat transfer coefficient.

3.58 The mass velocity on the shell side (lb/h·ft²) is most nearly:

a. 564,700
b. 250,000
c. 350,000
d. 850,000

3.59 The shell side heat transfer coefficient [Btu/(h·ft^2·°F)] is most nearly:

a. 900
b. 1000
c. 770
d. 850

3.60 For a countercurrent heat exchanger, inside and outside heat transfer coefficients h_i and h_o have been computed as 700 and 250 Btu/h·ft^2·°F, respectively. By experience, it is known that a dirt resistance, $R_{di} = 0.001$, will deposit annually on the inside of the tube while $R_{do} = 0.0015$ will deposit on the outside of the tube. The tubes are 1" OD and 0.902" ID. The thermal conductivity k of metal is 26 Btu/h·ft^2·°F/ft. The overall heat transfer coefficient [Btu/(h·ft^2·°F)] on the basis of which the heat transfer surface should be calculated so that the exchanger need be cleaned only once a year is nearly:

a. 100
b. 120
c. 172
d. 150

Problems 3.61–3.63

Overall heat transfer coefficients based on outside area were determined in a series of experiments for the condensation of steam on the outside of a dirty and a clean tube with water flowing through the tubes at various velocities. Wilson plots of the two sets of data are plotted as $1/U_0$ vs $1/u^{0.8}$ in Exhibit 3.61. Assume constant steam film coefficient on the outside of the tubes. Tubes are admiralty-metal, 1" OD and 0.902" ID. Thermal conductivity of tube metal = 63 Btu/h·ft^2·°F/ft.

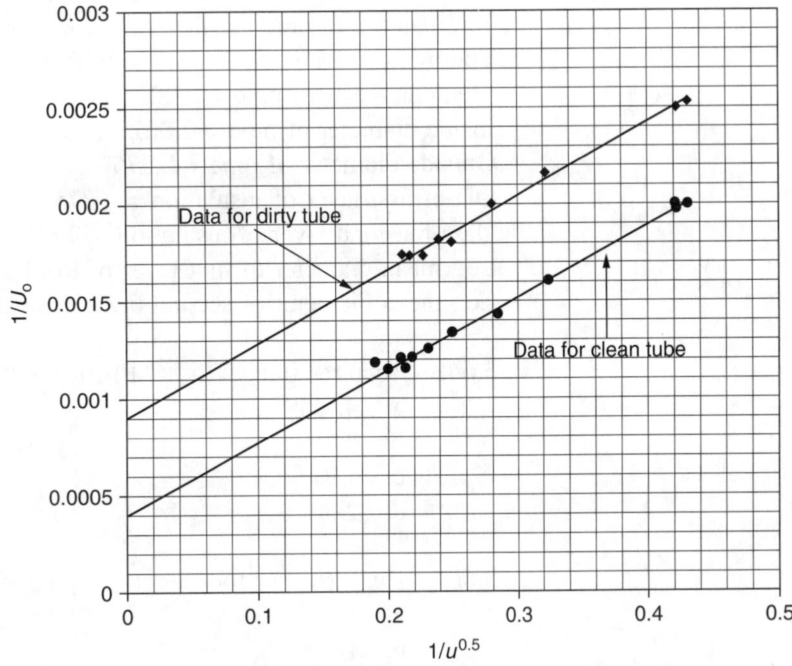

Exhibit 3.61 Wilson plot of condenser data for Problem 3.61

3.61 The waterside heat transfer coefficient [Btu/(h·ft^2·°F)] based on the inside area of the tubes at a velocity of 1 ft/s is closest to:
a. 246
b. 263
c. 252
d. 292

3.62 The condensation coefficient [Btu/(h·ft^2·°F)] for steam based on the outside area of the tube is nearest to:
a. 3012
b. 2750
c. 2500
d. 2000

3.63 The scale deposit coefficient h_d [Btu/(h·ft^2·°F)] based on the inside area of the tube is most nearly:
a. 2700
b. 2500
c. 2250
d. 2130

Problems 3.64–3.67

A horizontal, standard 2" pipe carrying superheated steam is insulated with the following:

First layer: 1.25"-thick diatomaceous earth + asbestos, $k = 0.058$ Btu/(h·ft^2·°F/ft)
Second layer: 2.5"-thick laminated asbestos felt, $k = 0.042$ Btu/(h·ft^2·°F/ft)

Other data:
Temperature of surroundings = 80°F
Average temperature of steam pipe = 900°F
Temperature of outer surface of second layer = 120°F
Thermal conductivity of steel = 26 Btu/(h·ft^2·°F/ft)
Mean diameter of pipe = 2.222"
Inside diameter of pipe = 2.067"
Outside diameter of pipe = 2.375"
OD of first layer of insulation = 4.875"
OD of second layer of insulation = 9.87"
Log mean diameter of first layer of insulation = 3.48"
Log mean diameter of second layer of insulation = 7.07"

3.64 Total resistance [(h·ft^2·°F)/Btu] of the two layers of insulation is closest to:
a. 4.651
b. 2.68
c. 1.97
d. 2.33

3.65 Heat loss per foot length of pipe [Btu/(h·linear ft of pipe)] is nearest to:
a. 162
b. 181
c. 168
d. 152

3.66 Interface temperature (°F) between the two insulation layers is nearest to:

a. 569
b. 352
c. 446
d. 331

3.67 The radiation heat transfer coefficient [Btu/(h·ft²·°F)] is closest to:

a. 0.87
b. 1.16
c. 1.0
d. 1.2

Hint: Kern (*Process Heat Transfer*) gives the following equation for convective heat transfer coefficient h_c from a horizontal pipe surface:

$$h_c = \left(\frac{\Delta t}{d_o}\right)^{0.25}$$

where

h_c = convective heat transfer coefficient from the surface, Btu/h·ft²·°F
d_o = outside diameter of pipe surface, in.

3.68 A 2" IPS steel pipe carrying steam at 300°F passes through a duct of galvanized sheet iron, which is maintained at 70°F. The cross section of the duct is 1 ft × 1 ft. The duct is insulated on the outside. Emissivity of steel, $\in = 0.8$. Emissivity of galvanized sheet iron, $\in = 0.28$. The heat loss [Btu/(h·linear ft of pipe)] is most nearly:

a. 150
b. 130
c. 166
d. 185

3.69 A bare 2" steel pipe carrying steam at 325°F passes through a room, which is at 70°F. The percent decrease in radiation that occurs if the bare pipe is coated with aluminum paint is nearest to:

a. 50.2
b. 45.3
c. 55.7
d. 60.1

Data: Emissivity of bare pipe = 0.8
Emissivity of painted pipe = 0.35

3.70 Oil is cooled at the rate of 45,300 lb/h in a countercurrent exchanger with cooling water flowing at the rate of 10,600 lb/h. The heat capacity of oil can be taken as 0.452 Btu/lb·°F and that of water as 1 Btu/lb·°F. The overall heat transfer coefficient is 53 Btu/h·ft²·°F. The effective area of the exchanger is 325 ft². The outlet temperature of oil (°F) is most nearly:

a. 150
b. 160
c. 120
d. 175

MASS TRANSFER

3.71 Acetic acid (A) is diffusing across a 0.00328-ft-thick film of non-diffusing water (B) solution at 77°F. The concentrations on opposite sides of the film are 10 and 3 wt% acid. The diffusivity of acetic acid in water is 4.8×10^{-5} ft²/h. The other data pertaining to the two sides of the film are given in the following table:

	10% Concentration Side	3% Concentration Side
Mol fraction acid	$x_{A1} = 0.0323$	$x_{A2} = 0.0092$
Average mol weight	19.38	18.41
Density, lb/ft³	63.18	62.56
ρ/M lb mol/ft³	3.26	3.398

Under these conditions, the rate of diffusion of acetic acid (lb mol/h·ft²) is most nearly:

a. 1.15×10^{-3}
b. 2.9×10^{-3}
c. 2.19×10^{-9}
d. 2.2×10^{-5}

3.72 Diffusivity of n-butane vapor in nitrogen at $t = 25°C$ and 1 atm pressure is 0.96×10^{-5} cm²/s. Its diffusivity at 60°C and 2 atm pressure, in ft²/h, will be most nearly:

a. 1.18×10^{-5}
b. 2.26×10^{-5}
c. 2.3×10^{-5}
d. 1.36×10^{-5}

3.73 Hydrogen gas at 17°C and 2 atm is flowing through a pipe made of unvulcanized neoprene rubber with ID = 25 mm and OD = 50 mm. The solubility of hydrogen gas in neoprene at 17°C is 0.051 m³(STP) per m³ solid at 1 atm. The diffusivity of H_2 at 17°C and 1 atm is 1.03×10^{-10} m²/s. The partial pressure on the outside of the neoprene pipe is zero. The steady state hydrogen loss, in kg mol/s per meter of pipe length, is most nearly:

a. 4.26×10^{-9}
b. 4.17×10^{-12}
c. 6.02×10^{-12}
d. 3.02×10^{-12}

3.74 The vapor pressure data at 200°F for the system hexane-octane are given below.

Hexane, 197.3 kPa; Octane, 37.1 kPa

Assuming the system obeys Raoult's law, the vapor composition of hexane (mol fraction) at a total pressure of 101.32 kPa and at 200°F will be most nearly:

a. 0.409
b. 0.781
c. 0.56
d. 0.65

Problems 3.75–3.77

i-Butanol forms a minimum-boiling-point azeotrope. The *t-x* diagram for this system is shown in Exhibit 3.75.

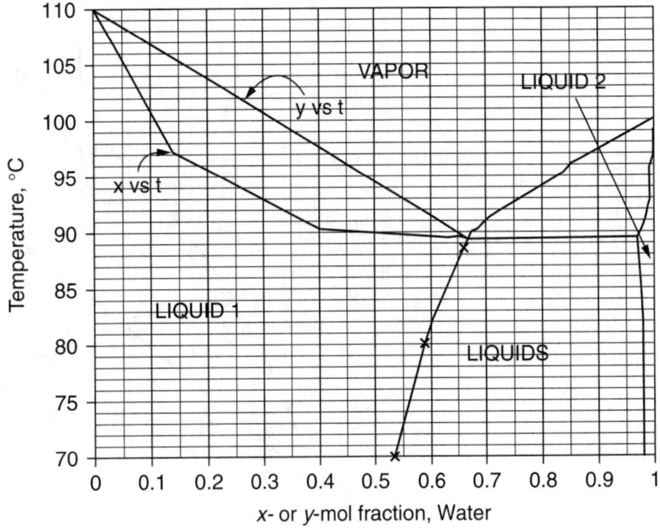

Exhibit 3.75 *t-x* diagram for i-butanol–water system (Problems 3.75–3.77)

3.75 The i-butanol concentration (wt %) of the azeotrope is most nearly:

a. 61.7
b. 66.04
c. 69.1
d. 56.3

3.76 10,000 lb of a vapor mixture of 23 mol % i-butanol and 77 mol % water vapor at 105°C is condensed at a constant pressure of 1 atm and subcooled to a temperature of 80°C. After phase separation, the butanol concentration (wt %) of the butanol-rich layer will most nearly be:

a. 80.66
b. 95
c. 75.62
d. 44.9

3.77 A vapor mixture containing 50 mol % i-butanol and 50 mol % water vapor at 110°C and 1 atm pressure will have a dew point temperature most nearly:

a. 95
b. 100
c. 92.8
d. 108

3.78 Exhibit 3.78 is an equilibrium diagram showing vapor-liquid compositions of heptane at 1 atm pressure for the system heptane-ethyl benzene.

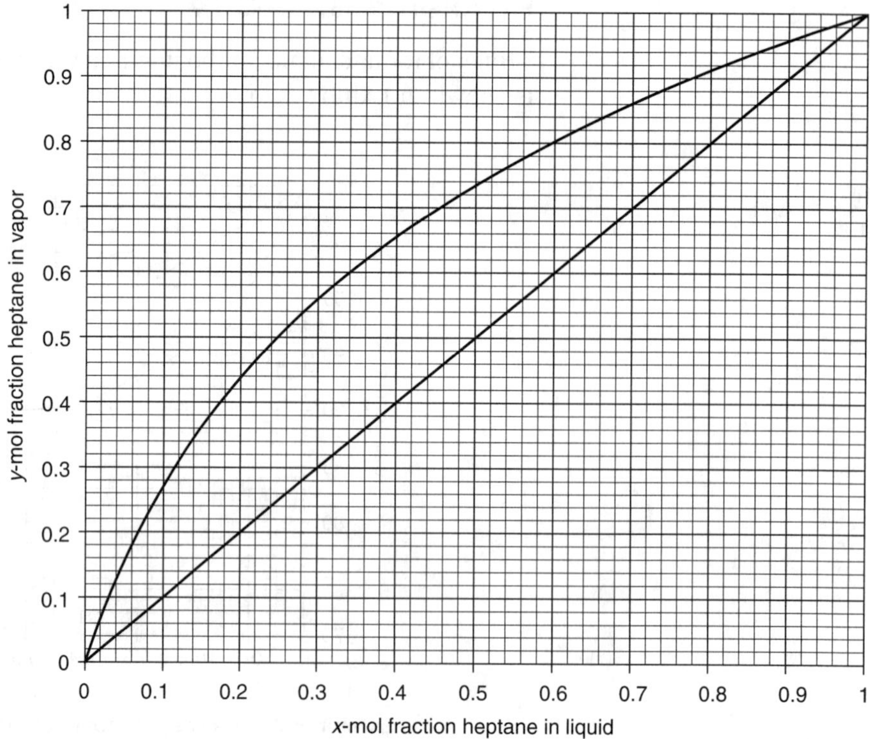

Exhibit 3.78 Equilibrium diagram for system heptane-ethyl benzene for Problem 3.78

A feed mixture containing 40 mol % heptane and 60 mol % ethyl benzene is to be fractionated at 1 atm pressure to produce a distillate containing 97 mol % heptane and a residue containing 98 mol % ethyl benzene.

If the feed to the column is saturated liquid and the reflux is at its bubble point, the minimum reflux ratio required is nearest to:

a. 0.53
b. 1.156
c. 1.34
d. 1.73

Problems 3.79–3.82

The specifications of the compositions of feed, distillate, and bottoms for the separation of the components by fractional distillation are given in the following table. The components are listed in the order of decreasing K values. The material balance is also given in the table.

	Feed		Distillate		Bottoms			
Component	**mol**	**mol fraction**	**mol**	**mol fraction**	**mol**	**mol fraction**	α_i **124°C**	α_i **165°C**
Phenol	35	0.35	31.5	0.953	3.5	0.0524	1.25	1.25
o-Cresol	15	0.15	1.5	0.0455	13.5	0.2020	1.00	1.0
m-Cresol	30	0.30	0.05	0.0015	29.95	0.4470	0.69	0.72
Xylenols	15	0.15			15.00	0.2240	0.44	0.48
Heavies	5	0.05			5.00	0.0750	0.088	0.089
Total	100	1.00	33.05	1.00	66.95			

Temperature at the top of column = 124°C
Temperature at the bottom of the column = 165°C
Average temperature in the column =147°C

Relative volatility values α_i at distillate and average temperatures are given in the last two columns of the table.

The still pressure is 250 mm Hg abs.

3.79 The heavy key component is:
a. o-cresol
b. m-cresol
c. xylenol
d. phenol

3.80 The minimum number of theoretical plates required using Fenske equation is nearest to:
a. 9.6
b. 19.7
c. 13.3
d. 16.4

3.81 If the minimum reflux ratio calculated by the Underwood method is 5.3622, and the actual reflux ratio used is 10, the number of theoretical stages required from the Erbar-Maddox correlation is most nearly:
a. 13.4
b. 39.7
c. 26.3
d. 30.9

3.82 Using the total number of theoretical stages obtained in Problem 3.81, and using the Kirkbride equation, the theoretical feed stage will be:
a. 13
b. 7
c. 9
d. 15

3.83 A packed tower is to be designed to treat 30,000 cu ft of gas per hour to remove NH_3 from it. The ammonia content of the gas is 5% by volume. Ammonia-free water will be used as an absorbent. The temperature is 68°F, and the pressure is 15 psia. The ratio of liquid to gas flow rates is 1, and $1^1/_2$" Ceramic Intalox saddles (F_p = 52) will be used as packing.

The percent flood that the column will operate at is nearest to:
a. 50
b. 55
c. 65
d. 70

3.84 A sieve tray column is to be designed to effect an aromatics separation. The following data are available pertaining to the operation of the fractionating column.

	Vapor	Liquid
Flow rate, lb/h	79,000	121,500
Vol. flow rate, cfs/(gpm)	76.13	(327)
Density, lb/ft^3	0.288	46.4
Surface tension = 18 dyn/cm		

tray spacing = 24", weir height = 2", hole i.d. = $^1/_4$", hole area/active area = 15%

Area of down-comer = 10% of tray area.

Also, assume waste = 7.5% of total tray cross section.

If the column is to operate at 80% flood, the diameter (ft) of the column for the required flows will be nearest to:

a. 5
b. 6
c. 6.5
d. 7

Problems 3.85–3.86

A single-effect evaporator is to concentrate 30,000 lb/h of caustic solution from 10% to 50% concentration. The evaporator is to be supplied with 30-psia steam. The feed is at 100°F. It operates at a vacuum of 26" Hg. Some additional data are given below.

Overall coefficient of heat transfer = 450 Btu/h·ft^2·°F
Boiling point of caustic solution = 198°F
Specific heat of steam, C_P = 0.46 Btu/lb·°F
Cooling water temperature = 85°F
Temperature of water leaving the condenser = 120°F

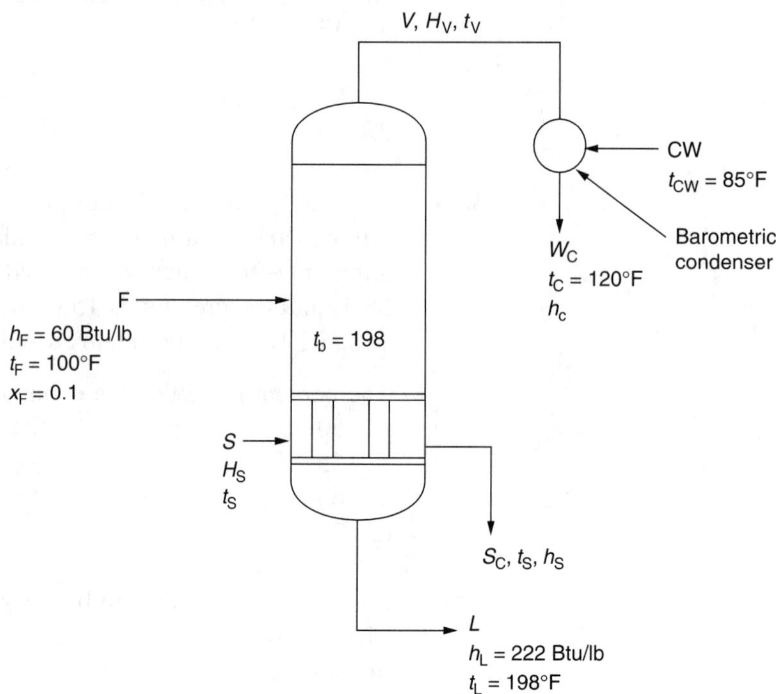

Exhibit 3.85 Sketch of single-effect evaporator for Problems 3.85 and 3.86.

3.85 The steam economy lb evaporated per lb of steam, is nearest to:
a. 1.0
b. 0.5
c. 0.837
d. 0.75

3.86 Cooling water required (gpm) is near to:
a. 1200
b. 1100
c. 1500
d. 1300

3.87 An insulated rotary dryer is to be designed to dry a crystalline material from 30% to 0.5% moisture content. The drier is to be 5 ft in diameter. The moisture in the material can be treated as unbound moisture. Air at 80°F (dry bulb) and 65°F (wet bulb) temperature will be heated to 240°F and fed to the drier countercurrent to the flow of the wet material. The air will leave the dryer at 110°F dry bulb. The wet solids will enter the dryer at 80°F and are to be discharged at 140°F. The specific heat of the dry solids is 0.3 Btu/(lb·°F). The volumetric heat transfer coefficient for rotary dryers is evaluated by the following equation proposed by Friedman and Marshall:

$$Ua = \frac{10G^{0.16}}{D}$$

where
Ua = volumetric coefficient of heat transfer, Btu/h·ft^3·°F

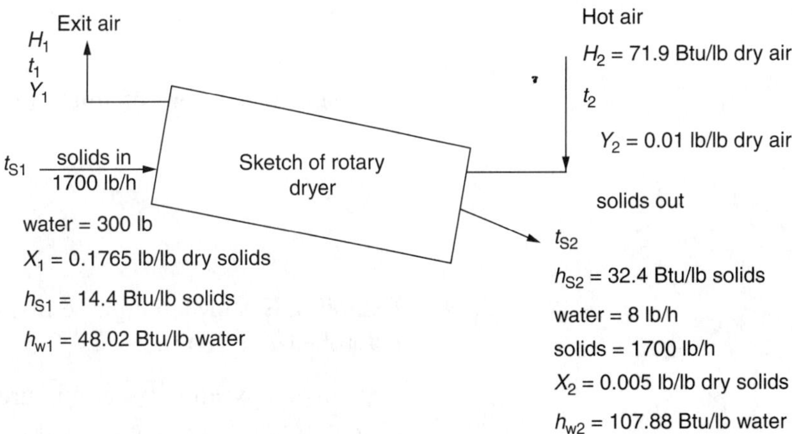

Exhibit 3.87 Material balance for a rotary dryer

To effect the required drying, the dry air rate, in lb/h, to the rotary dryer will be nearest to:
a. 5000
b. 7000
c. 9850
d. 11,000

3.88 1000 lb of a solid containing 50% solute and 50% inerts is to be extracted with solvent C in a two-stage crosscurrent extraction using 3000 lb of solvent each time. In each stage, the extracted solids are screw-pressed. The pressed solid contains 1.2 lb of solution per lb of inerts. Exhibit 3.88 provides a triangular diagram for solving the problem.

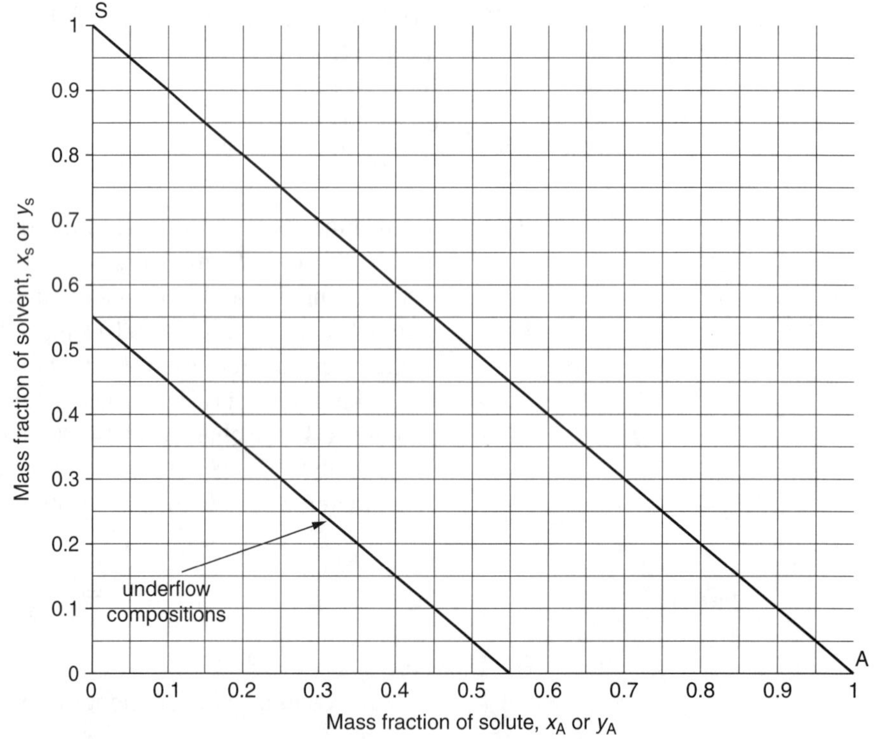

Exhibit 3.88 Isosceles triangular diagram for the solution to Problem 3.88

The percent recovery of solute in the two-stage process is nearest to:

a. 98
b. 99
c. 90
d. 93

3.89 Exhibit 3.89 shows the phase diagram for the pyridine (C)-water (A)-chlorobenzene (B) system.

2000 lb of a pyridine-water mixture free of chlorobenzene and containing 50% pyridine is to be extracted in a single-batch extraction with chlorobenzene as the solvent. The concentration of pyridine in the raffinate is to be reduced from 50% to 2% mass fraction. The solvent amount required, in lb/batch, is nearest to:

a. 20,400
b. 21,867
c. 23,206
d. 18,674

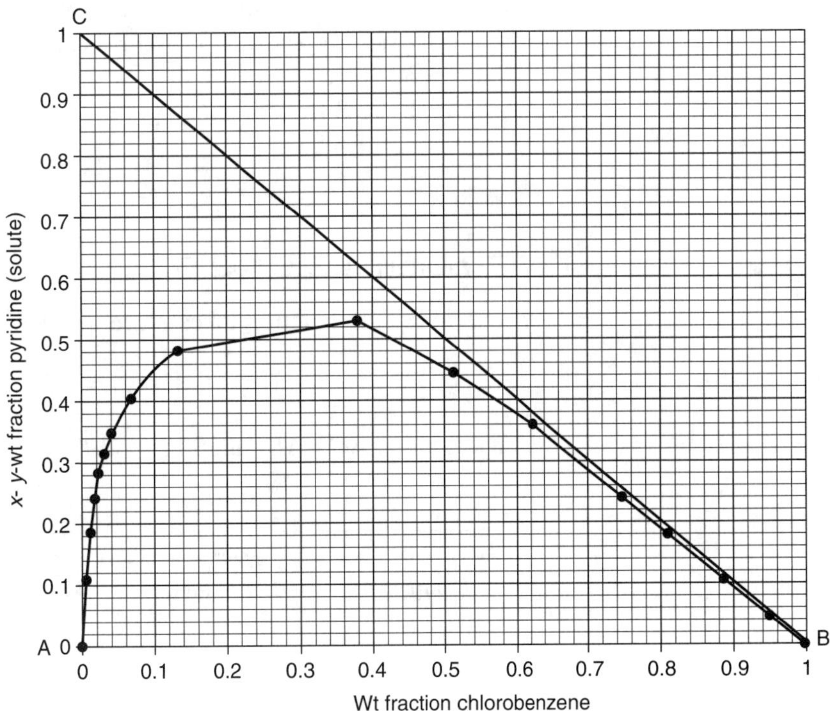

Exhibit 3.89 Pyridine-water-chlorobenzene system

3.90 Air enters a cooling tower at the bottom at a rate of 1400 lb/h·ft^2. Its dry-bulb temperature is 85°F and wet-bulb temperature is 75°F. The humidity of dry air entering the tower (lb of water vapor/lb dry air) is closest to:

a. 0.002
b. 0.005
c. 0.0168
d. 0.014

CHEMICAL REACTION ENGINEERING

Problems 3.91–3.93

The rate of a liquid-phase reaction $A + B \rightarrow$ Products is found to be independent of reactant concentrations. The reaction rate is 1.405 g mol/liter·min at 300 K. The feed concentrations of A and B are 5 g mol/liter. The feed-flow rate is 1.2 m^3/min, and the reactor operates at a temperature of 300 K. The activation energy of the reaction at 300 K is 53.2 kJ/mol.

3.91 The conversion attained in a continuous mixed flow reactor is most nearly:

a. 0.99
b. 0.7025
c. 0.75
d. 0.60

3.92 The value of the reaction rate constant (g mol/liter·min) at 310 K is nearest to:

a. 2.80
b. 6.36
c. 3.16
d. 5.6

3.93 With feed conditions as in Problem 3.91, the volume (m^3) of a plug flow reactor operating at 310 K will be most nearly:

a. 0.25
b. 3.34
c. 1.5
d. 2.6

Problems 3.94–3.97

The following reaction takes place at 704°C and 2 atm:

$$C_4H_{10}(g) = 2C_2H_4(g) + H_2(g)$$

The data on ΔG^o, ΔH^o at 298 K and C_p values are given in the following table:

Component	ΔG^o_{298} kcal/g mol	ΔH^o_{298} kcal/g mol	C_p kcal/g mol·K
C_4H_{10}	−3.75	−29.81	$C_p = 0.01178 + 4.268 \times 10^{-5}$ T
C_2H_4	16.28	12.5	$C_p = 0.0028 + 3.000 \times 10^{-5}$ T
H_2	0	0	$C_p = 0.0069 + 0.4 \times 10^{-5}$ T

3.94 The free energy (kcal/gmol) of the reaction at 25°C is nearest to:

a. 36.31
b. −36.31
c. 20.03
d. −20.3

3.95 The heat of reaction, in kcal/gmol, for this reaction at a temperature of 500°C is nearest to:

a. −60.57
b. 60.57
c. −29.93
d. 29.93

3.96 The equilibrium constant K for the reaction at a temperature of 704°C is closest to:

a. 6.5
b. 133
c. 303.7
d. 62.3

3.97 The equilibrium conversion (%) of n-butane at a temperature of 704°C is nearest to:

a. 99.56
b. 65.53
c. 80.21
d. 91.17

[Hints: When the component C_ps are expressed by equations of the type

$$C_p^o = \alpha + \beta T + \gamma T^2 + \cdots$$

One can evaluate net C_p^o for products and reactants as given below.

$$\Delta C_p^o = \Delta\alpha + \Delta\beta T + \Delta\gamma T^2 + \cdots$$

where

$\Delta\alpha = (\Sigma n\alpha)_{\text{products}} - (\Sigma n\alpha)_{\text{reactants}}$, with similar equations for $\Delta\beta$ and $\Delta\gamma$

Then various relations have been developed as follows:

Heat of reaction at any temperature,

$$\Delta H_T^o = I_H + \Delta\alpha T + \Delta\beta\left(\frac{1}{2}\right)T^2 + \Delta\gamma\left(\frac{1}{3}\right)T^3 + \cdots$$

The equilibrium constant, K, is related to the temperatures by the following equation:

$$\ln \text{K} = -\frac{I_H}{RT} + \frac{\Delta\alpha}{R} T \ln T + \frac{\Delta\beta}{R}\left(\frac{1}{2}\right)T + \frac{\Delta\gamma}{R}\left(\frac{1}{6}\right)T^2 + \cdots + \text{I}$$

where I_H is another constant that can be evaluated if one value of K is known. Also, the following relation gives free energy of a reaction at any temperature T

$$\Delta G_T^o = I_H + I_G T - \Delta\alpha T - \frac{1}{2}\Delta\beta T^2 - \frac{1}{6}\Delta\gamma T^3 + \cdots$$

where

$I_G = -(\text{IR})$]

3.98 A gas-phase reaction was carried out in a constant-volume batch reactor. Times for 50% conversion were determined at various concentrations and at a temperature of 110°C. The data are reported as follows:

C_{AO}	$\ln C_{AO}$	$t_{1/2}$	$\ln t_{1/2}$
0.01	–4.61	5	1.61
0.025	–3.69	2	0.69
0.05	–3	1	0
0.075	–2.59	0.67	–0.4
0.1	–2.3	0.5	–0.69

The order of the reaction is:

a. 3
b. 2
c. 1.5
d. 1

3.99 The hydrolysis of acetic anhydride is carried out in a series of four same-size mixed flow reactors. The reactors are operated at different temperatures: 10°C, 15°C, 25°C, and 40°C, respectively. The inlet composition is 1.5 lb mol/gal, and the flow rate is 30 gpm. The desired conversion is 95%. The reaction is first order. The rate constants are given in the following table:

Temperature, °C	10	15	25	40
k, min^{-1}	0.0567	0.0806	0.158	0.38

If all the reactors are maintained at a temperature of 15°C, the number of the same size reactors required is:

a. 5
b. 4
c. 3
d. 7

3.100 The Damköhler number for a second-order liquid-phase reaction carried out in a CSTR was found to be 6. The percent conversion for this reaction is nearest to:

a. 70
b. 60
c. 67
d. 82

3.101 A liquid-phase first-order reaction is carried out in a 750-gal CSTR. The reaction rate constant k is 0.3 min^{-1}. The feed rate to the reactor is 15 ft^3/m. The Damköhler number for this reaction is nearest to:

a. 3.1
b. 2.5
c. 2.0
d. 3.4

3.102 A liquid-phase reaction of the first order with a reaction rate constant $k = 0.3$ min^{-1} is carried out in two CSTRs in series, each with a 750-gal volume. The feed rate is 15 ft^3/min. The conversion attained will be most nearly:

a. 0.85
b. 0.89
c. 0.93
d. 0.97

3.103 A reaction has an activation energy of 82,324 J/g mol. The temperature range (°C) in which the rate of this reaction will double is most nearly:

a. 120–130
b. 100–110
c. 90–100
d. Any range

Problems 3.104–3.105

For consecutive reactions $A \xrightarrow{k_1} B \xrightarrow{k_2} C$, the values of the rate constants k_1 and k_2 are

$$k_1 = 0.35 \text{ h}^{-1},\ k_2 = 0.13 \text{ h}^{-1},\ C_{A0} = 5 \text{ lb mol/ft}^3,$$
$$C_{B0} = 0, \text{ and } C_C = 0 \text{ at } t = 0$$

[The rate equations for the two reactions and their solutions in the case of a batch reactor are

$$r_a = -\frac{dC}{dt} = k_1 C_A$$
$$r_b = -\frac{dC_B}{dt} = k_2 C_B - k_1 C_a$$

$$C_A = C_{A0} e^{-k_1 t}$$
$$C_B = C_{A0} \frac{k_1}{k_2 - k_1}(e^{-k_1 t} - e^{-k_2 t})$$
$$C_C = C_{A0}\left(1 - \frac{k_2}{k_2 - k_1} e^{-k_1 t} + \frac{k_1}{k_2 - k_1} e^{-k_2 t}\right)$$

For a continuous, mixed reactor, the following relations apply:

$$C_{A1} = \frac{C_{A0}}{1 + k_1\theta}$$

$$C_{B1} = \frac{k_1 C_{A0}\theta}{(1 + k_1\theta)(1 + k_2\theta)}$$

where
θ = the residence time in the vessel.]

3.104 The maximum concentration (lb mol/ft^3) attained by B when operating as a batch reactor is nearest to:
a. 2.64
b. 2.53
c. 2.79
d. 2.94

3.105 For a single-stage continuous, mixed reactor, the maximum concentration (lb mol/ft^3) of B in the effluent stream is closest to:
a. 1.62
b. 1.43
c. 1.925
d. 1.74

3.106 A decomposes to products B, C, and D according to the following reactions:

$$A \xrightarrow{k_1} B \qquad A \xrightarrow{k_2} C \qquad A \xrightarrow{k_3} D$$

The values of the reaction specific rates are $k_1 = 0.36\ \text{h}^{-1}$, $k_2 = 0.12\ \text{h}^{-1}$, and $k_3 = 0.1\ \text{h}^{-1}$.

At the start of the reaction, $N_{a0} = 100$ g mol, $N_{b0} = N_{c0} = N_{d0} = 0$. After 1 hour of reaction, the g mol of D in the reaction mixture will be:

a. 7.59
b. 9.1
c. 22.8
d. 17.1

Problems 3.107–3.110

A solution containing a reactive component, with an initial concentration, C_{A0} of 0.5 lb mol/ft^3 is to be treated in different types of reactors. The feed rate for a continuous-flow operation is to be 25 ft^3/h. The reaction rate data for the decomposition of A are as follows:

C_A, lb mol/ft^3	$-r_A$, lb mol/(h·ft^3)	$-1/r_A$
0.5	0.85	1.1765
0.4	0.53	1.8868
0.3	0.31	3.2258
0.2	0.18	5.5556
0.1	0.081	12.3457
0.05	0.04	25.00

3.107 If the filling and draining time per batch is negligible, the number of batches that can be processed per day in a batch reactor is most nearly:

a. 12
b. 20
c. 15
d. 10

3.108 If the reaction is carried out in a continuous, mixed flow reactor and the processing rate is 25 ft^3/h, the volume (ft^3) of the reactor required to realize 90% conversion is nearest to:

a. 100.4
b. 281.3
c. 220.8
d. 320.6

3.109 If each vessel has a volume of 50 ft^3, the percent conversion attained with two continuous, mixed flow reactors is nearly:

a. 90.2
b. 75.4
c. 85.2
d. 95.1

3.110 If a plug flow reactor is used, the volume (ft^3) of the reactor to effect 90% conversion is closest to:

a. 60.0
b. 100
c. 90
d. 95.1

PLANT DESIGN AND OPERATIONS

3.111 Today's cost of a carbon steel shell-and-tube heat exchanger with 3000 ft^2 of heating surface and designed for 150 psig/300°F is $50,000. If the annual inflation rate is 3%, the approximate installed cost ($) of a similar unit with 9019 ft^2 of heating surface area three years from now will be most nearly:

a. 395,750
b. 432,440
c. 354,100
d. 346,400

3.112 The installed cost of a machine is $20,000. Its salvage value after 10 years of useful life is $4000. Interest rate is 10%, and the annual maintenance cost is $1000. The annual cost ($) of the machine is most nearly:

a. 3004
b. 3618
c. 4004
d. 4506

3.113 The estimated annual fixed costs and the costs incurred due to heat loss for a certain insulation as a function of insulation thickness are plotted in Exhibit 3.113a.

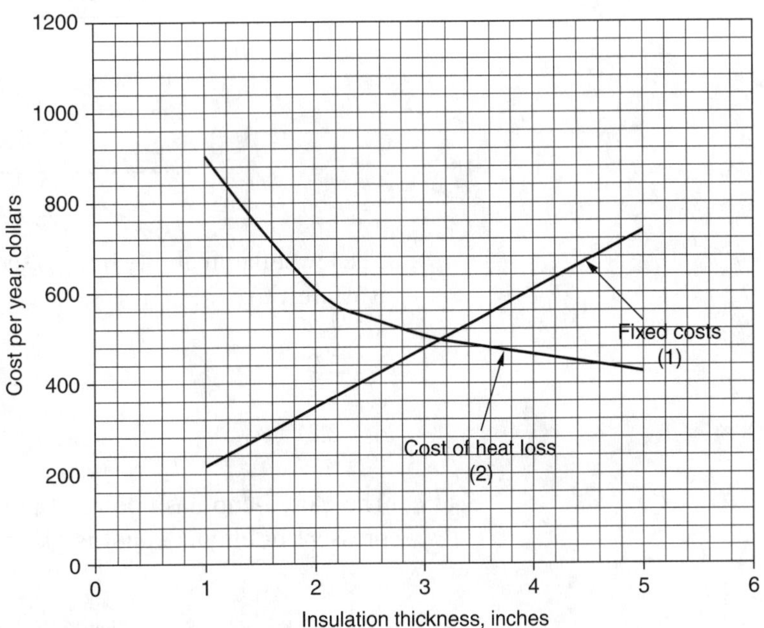

Exhibit 3.113a Fixed costs and cost due to heat loss

These data indicate an optimum insulation thickness, in inches, of most nearly:

a. 3.25
b. 5.2
c. 2.25
d. 4

3.114 The annual direct-production costs for a plant operating at 70% capacity are \$350,000. Total annual fixed charges, overhead costs, and general expenses are \$250,000. If total sales are \$700,000, and the product sells at \$50 per unit, the break-even point, in units of production, is nearest to:

a. 12,000
b. 9000
c. 10,000
d. 11,000

3.115 A clarifier is used to remove suspended solid particles of size 30 μm (specific gravity = 2.5 using type 1 sedimentation settling). The viscosity of the liquid is 1.002 cP. The depth (ft) to which the particles will settle in two hours will be most nearly:

a. 15
b. 6
c. 10
d. 12

3.116 A combustible mixture of methane, hexane, and ethylene in air has composition and lower flammability limits at 298 K and 1 atm as given in the following table:

Component	Mol %	LFL %
Methane	2.1	5.00
Hexane	0.9	1.1
Ethylene	0.6	0.6
Air	96.4	

The lower flammability limit (% combustibles) of the combustible mixture is nearest to:

a. 1.3
b. 2.9
c. 1.6
d. 2.3

3.117 The closed-cup flash point of pure ethyl alcohol is 13°C. The vapor pressure of ethyl alcohol is given by the following Antoine equation:

$$\log p = A - \frac{B}{T + C}$$

where
T = temperature, °C and
p = vapor pressure, mm Hg

The values of the constants are $A = 8.112$, $B = 1592.864$, and $C = 226.184$.

The closed-cup flash point (°C) of a solution of 90 wt % ethyl alcohol and 10 wt % water will be most nearly:

a. 13.8
b. 15.6
c. 17.7
d. 19.4

3.118 A conservation vent for a tank is to be designed for the storage of a flammable liquid under the following conditions:

Maximum liquid temperature = 115°F
Minimum liquid temperature = 100°F
Maximum vapor space temperature = 155°F
Minimum vapor temperature = 100°F
Liquid vapor pressure at 115°F = 4.5 psia
Liquid vapor pressure at 100°F = 3.2 psia
Elevation of tank from sea level = 3000 ft

Under these conditions the estimated set pressure (psig) for the breather vent will be most nearly:

a. 4.5
b. 2.3
c. 3.2
d. 3.9

[Additional information: An estimate of the breather vent set pressure may be made by the following relation:

$$P_s = (P_a - P_{\min})\frac{T_2}{T_1} + P_{\max} - P_a$$

$$P_a = 14.696\left(1 - \frac{0.0065\Delta Z}{288}\right)^{5.26433}$$

where
P_a = local barometric pressure, psia
P_s = set pressure of safety device, psig
ΔZ = elevation of location above sea level, m
$P_{\max}$ = vapor pressure of liquid at maximum liquid surface temperature, psia
$P_{\min}$ = vapor pressure of liquid at minimum liquid surface temperature, psia
T_1 = minimum vapor space temperature, °R
T_2 = maximum vapor space temperature, °R]

Problems 3.119–3.124

A carbon-steel tank is planned to be located as shown in Exhibit 3.119. Its main purpose is to store toluene. A relief valve is to be sized for protecting the tank against overpressure from fire.

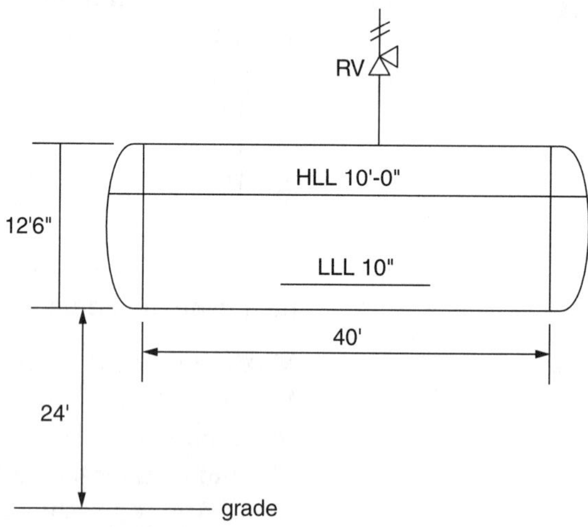

Exhibit 3.119 Diagram for Problems 3.119–3.124

The other pertinent data are as follows:

Design pressure = 190 psig
Set pressure = 190 psig
Back pressure = 0 psig
Insulation for heat conservation
No water spray, no drainage facility
ASME flanged and dished heads

Properties of liquid:

Boiling point of liquid = 383.6 K
Latent heat of vaporization = 14,400 Btu/lb mol@BP
Critical temperature = 593.9 K
Critical pressure = 40.3 atm

Heat capacity, C_P of toluene is given by the expression

$$C_P = 0.09418 + 0.03838 \times 10^{-3} T - 0.2786 \times 10^{-6} T^2 + 0.08033 \times 10^{-9} T^3 \ \frac{\text{kJ}}{\text{mol} \cdot {}^\circ\text{C}}$$

where T is in °C.

The vapor pressures of toluene (mm Hg) as a function of temperature are given by

$$\log p = A - \frac{B}{T + C}$$

The values of the Antoine constants A, B, and C in the above equation are as follows:

$A = 6.95805$, $B = 1346.773$, and $C = 219.63$

3.119 The relief load, (Btu/h) is most nearly:
a. 11.13×10^6
b. 12.27×10^6
c. 11.8×10^6
d. 10.32×10^6

3.120 The relieving temperature (°C) is nearest to:
a. 110
b. 154
c. 203
d. 252

3.121 The relief load (lb/h) is most nearly:
a. 71,255
b. 120,300
c. 108,690
d. 101,176

3.122 The heat capacity ratio for toluene at the relieving temperature is most nearly:
a. 1.1054
b. 1.3023
c. 1.2079
d. 1.2653

3.123 The critical flow pressure, P_{CF} (psia) is most nearly:
a. 238.1
b. 87.7
c. 223.1
d. 195.8

3.124 The orifice size of a conventional safety valve to be selected for the required service (relief load of 108,691 lb/h) is nearest to:
a. L
b. K
c. M
d. N

3.125 The saponification of fatty acids is carried out in kettles using 12.6% to 14% NaOH solution. Heat is supplied by direct steam. The kettle is kept boiling for four hours. After four hours, NaCl is added, and boiling is continued another four hours. Soap separates as the upper layer, and glycerin and NaCl separate as the lower layer. The most economic construction material to carry out this operation is:
a. Carbon steel
b. Stainless steel 304
c. Glass lined steel
d. Stainless steel 316

3.126 Of the following four methods, the one that will actually increase the galvanic corrosion of a more active metal is:

a. Using combination of two metals as close as possible in galvanic series
b. Coupling two widely separated metals in the galvanic series
c. Protective oxide films
d. Insulating the two metals from each other.

3.127 A chemical company produces chlorinated and fluorinated chemicals. Reactor pressure vessels are constructed of alloy C-276 (Ni = 57%, Cr = 16%, Mo = 16%, and Fe = 5%). Because of excessive corrosion, they have to be replaced every 12 to 14 months. The corrosion rate of C-276 in 10% H_2SO_4 + 1% HCl solution at 90°C is 0.041 inches per year, while that of alloy 59 is 0.003 inches per year. The process involves chemicals such as hydrocarbons, ammonium fluoride, sulfuric acid, and others. A suitable new construction material for the reactor to have a more prolonged life than the reactors made of alloy C-276 would be:

a. Glass lined steel
b. Tantalum clad vessel
c. Alloy 59 (Ni = 59%, Cr = 23%, Mo = 16%, Fe = 1%)
d. SS 317L

3.128 A control system has transfer functions as shown in Exhibit 3.128.

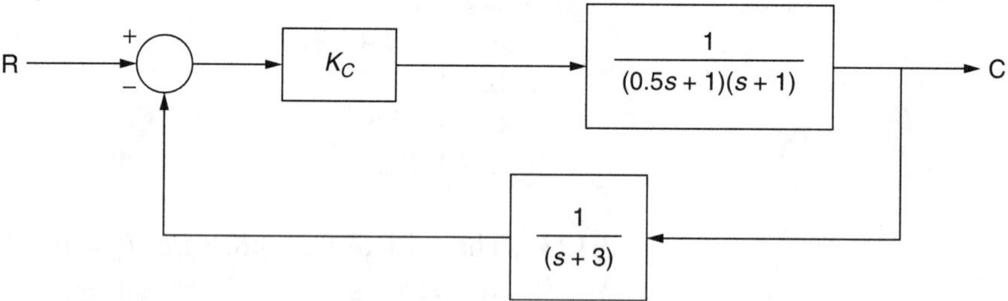

Exhibit 3.128 Diagram for Problem 3.128

By applying the Routh test, only the following can be said about this control system:

a. The system is stable if the value of $K_C < 10$.
b. The system is oscillatory if $6 < K_C < 10$.
c. The system is stable as well as non-oscillatory if $0 < K_C < 6$.
d. The system is unstable at all values of K_C.

3.129 The block diagram shown in Exhibit 3.129 represents the overall transfer function for two CSTRs in series with a feedback controller.

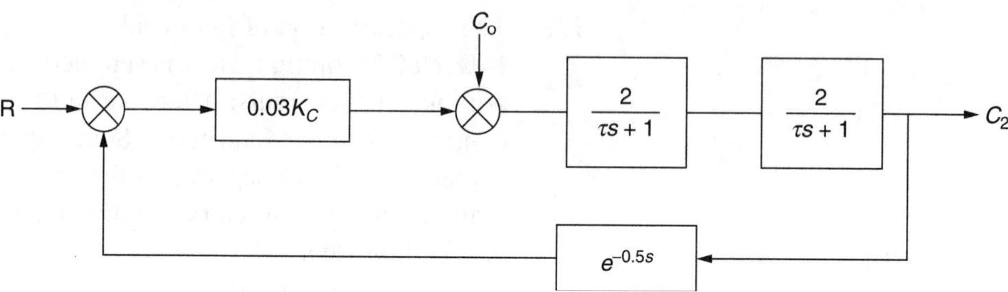

Exhibit 3.129 Diagram for Problem 3.129

If a unit step function is applied to the set point, the offset is nearest to:

a. $\frac{1}{1+0.12K_C}$

b. $\frac{0.12K_C}{1+0.12K_C}$

c. $0.12K_C$

d. 0

3.130 If V is the volumetric flow rate, and m is the output signal of a magnetic flow meter, the variation of m with V is given by:

a. $m \propto \frac{1}{V}$

b. $m \propto V$

c. $m \propto V^2$

d. $m \propto \sqrt{V}$

CHAPTER 4

Solutions to Sample Exams

OUTLINE

MORNING EXAM (CHAPTER 1) SOLUTIONS

1.1 **c.** Prepare a table of the combustion products and the theoretical amount of oxygen required for complete combustion.

Basis: 100 lb of fuel

Component	Fuel lb/(100 lb fuel)	CO_2 lb mol	SO_2 lb mol	H_2O lb mol	O_2 Used lb mol
C	87.94	7.322			7.322
H	10.57			5.233	2.617
S	0.84		0.026		0.026
O	0.65				–0.020
					9.945

$$
\begin{aligned}
\text{With 20\% excess, the oxygen used} &= 9.945(1.2) \\
&= 11.934 \text{ lb mol} \\
&= 11.934(32) \\
&= 381.9 \text{ lb/100 lb of fuel}
\end{aligned}
$$

1.2 **d.** From Orsat analysis, it is clear that the combustion is complete. The products of the combustion are given in the following table.

Basis: 100 mol of natural gas

Component	CO_2 lb mol	H_2O lb mol	N_2 lb mol	O_2 lb mol Consumed
CH_4	97.3	194.6		97.3 + 194.6/2 = 194.6
N_2			2.3	
CO_2	0.4			
Total	97.7			

Flue gas from 100 lb mol of natural gas:

$$N_2 = \frac{85.94}{8.8} \times 97.7 = 954.13 \text{ lb mol}$$

$$CO_2 = 97.7 \text{ lb mol}$$

$$N_2 \text{ in flue gas} = \frac{85.94}{8.8} \times 97.7 = 954.1 \text{ lb mol}$$

$$N_2 \text{ from air} = 954.1 - 2.3 = 951.8 \text{ lb mol}$$

$$O_2 \text{ in flue gas} = \frac{5.26}{8.8} \times 97.7 = 58.4 \text{ lb mol}$$

Therefore, O_2 from air = 951.8(0.21/0.79) = 253.0 lb mol
O_2 consumed in combustion = 253.0 – 58.4 = 194.6 lb mol (check)

Then, percent excess $O_2 = \frac{58.4}{194.6} \times 100 = 30\%$

1.3 **a.**

$$\text{24-hour supply of gas} = 24(3) = 72 \text{ ft}^3 \text{ (at 75°F and 772 mm Hg pressure)}$$

$$\text{Temperature} = 460 + 75 = 535°\text{R}$$

$$\text{Volume at standard conditions} = 72 \times \frac{492}{535} \times \frac{772}{760}$$

$$= 67.26 \text{ ft}^3 \text{ (at 32°F and 760 mm Hg)}$$

$$\text{lb of acetylene needed} = \frac{67.26 \text{ ft}^3}{359 \text{ ft}^3\text{/lb mol}} \times \frac{26.04 \text{ lb}}{\text{lb mol}} = 4.88 \text{ lb}$$

Therefore, calcium carbide needed $\frac{4.88}{26.04} \times 64.1 = 12.01$ lb.

1.4 **b.** $CaCO_3 \rightarrow CaO + CO_2$
100 56 44
Basis: 100 lb of $CaCO_3$

The reaction is 95% complete.

$CaCO_3$ decomposed = 95 lb per 100 lb $CaCO_3$

$$CO_2 \text{ produced} = \frac{44}{100} \times 95 = 41.8 \text{ lb}$$

Therefore, yield of CO_2 = 41.8/100 = 0.418 lb/lb limestone

1.5 **a.** Basis: 100 mol of CH_4

Component	Feed, lb mol	Product	Product Quantity, lb mol	O_2 Used, lb mol
CH_4	90	CO_2	90	90
N_2	10	H_2O	180	90

O_2 actually used = 1.25(180) = 225 lb mol
N_2 in flue gases = 10 + (0.79/0.21)(225) = 856.43 lb mol
CO_2 in boiler flue gas = 90 lb mol

Using N_2 as a tie component,

CO_2 in absorber exit gas = (1.1/93.9)(856.43) = 10.03 lb mol
CO_2 absorbed = 90 – 10.03 = 79.97 lb mol

$$\text{Percent CO}_2\text{ absorbed} = \frac{79.97}{90}(100) = 88.86\%$$

1.6 **c.** The steady-state energy balance with condensing steam as the system reduces to

$$Q = \hat{H}_O - \hat{H}_I = M(\hat{H}_O - \hat{H}_I)$$

From the SI steam tables,

Enthalpy of saturated steam at 150°C = 2745.4 kJ/kg.

Since the pressure drop on the condensing-steam side is very small, we can use the properties of steam at 150°C for the outlet stream without introducing any substantial error.

Outlet liquid enthalpy at 150°C = 632.1 kJ/kg
Outlet vapor enthalpy at 150°C = 2745.4 kJ/kg
Outlet mixed-stream enthalpy = 0.5(2745.4 + 632.1) = 1688.8 kJ/kg

Substituting these values into the energy-balance equation,

$$Q = 2500(1688.8 - 2754.4) = -2{,}641{,}500 \text{ kJ/min}$$

(The minus sign indicates that heat leaves the system.)

$$= \left(2.6415 \times 10^6 \frac{\text{kJ}}{\text{min}}\right)\left(\frac{1}{1.0544 \text{ kJ/Btu}}\right)$$

$$= -2.505 \times 10^6 \text{ Btu/min} = -2.505 \text{ MMBtu/min}$$

1.7 **a.** System: Take the reaction mass in the reactor as the system. The energy-balance equation gives

$$U_E - U_B = \overset{(2)+(1)}{\Sigma(H + PE + KE)_i} - \overset{(2)+(1)}{\Sigma(H + PE + KE)_o} + \Sigma Q + \overset{(3)}{W}$$

1. No kinetic energy and potential energy changes.
2. No mass in and out.
3. No work involved.

The energy-balance equation reduces to

$$U_E - U_B = \Sigma Q = Q_s - Q_r - Q_l$$

where
Q_s = Heat supplied by steam, Btu
Q_r = Heat due to endothermic reaction, Btu
Q_l = Heat loss

For liquids, $(U_E - U_B) \approx (H_E - H_B)$. Therefore, by replacing $(U_E - U_B)$ with $(H_E - H_B)$, we get

$$H_E - H_B = Q_s - Q_r - Q_l$$
$$H_E - H_B = \Delta H = Q_s - Q_r - Q_l$$
$$\Delta H = 500(0.8)(212-68) = 57{,}600 \text{ Btu/batch}$$
$$Q_r = \text{Heat of reaction} = 991(500) = 495{,}500 \text{ Btu/batch}$$
$$Q_l = \text{Heat loss} = 8000(2) = 16{,}000 \text{ Btu/batch (in 2 h)}$$

Therefore,

Q_s = Heat transferred to reaction mass from condensation of steam
= 57,600 + 495,500 + 16,000
= 569,100 Btu/batch

From the data table given in the problem statement,

λ_s at 450°F = (785.4 + 763.3)/2 = 774.35 ≈ 774.4 Btu/lb
Steam usage = m = (569,100)/774.4 = 735 lb

1.8 **b.** 2-Methyl butane, or isopentane, undergoes combustion according to the following equation:

$$C_5H_{12}(g) + 8O_2(g) = 5CO_2(g) + 6H_2O(l) \quad \Delta H_c = -843.216 \text{ kcal/g mol}$$
$$\Delta H_f \qquad \Delta H_f = 0 \qquad 5(-94.052) \qquad 6(-68.316)$$
$$\Delta H_r = \Delta H_c = \Sigma(\Delta H_f)_P - \Sigma(\Delta H_f)_R$$
$$-843.216 = 5(-94.052) + 6(-68.316) - (\Delta H_f)_{isopentane}$$

From which, $(\Delta H_f)_{isopentane} = -36.93$ kcal/g mol of $C_5H_{12}(g)$.

1.9 **b.** The steady-state energy-balance equation reduces to

$$0 = \hat{H}_I - \hat{H}_O - (KE)_O + W$$

or

$$W = \hat{H}_O - \hat{H}_I + (KE)_O$$

By substituting the given values,

$$W = 218.87 - 210.27 + \frac{200^2}{64.4(778)} = 9.4 \text{ Btu/lb}$$

Total theoretical work = (9.4)(5)/42.4 = 1.1085 ≈ 1.11 hp

Therefore,

$$\text{bhp} = 1.11/0.7 \approx 1.6$$

1.10 c.

$$\Delta H_V = T\Delta V\frac{dP}{dT}$$

For an approximate solution, we can write the equation as

$$\Delta H_V = T\Delta V\frac{\Delta P}{\Delta T}$$

$$\frac{\Delta P}{\Delta T} = \frac{32.624 - 27.134}{290 - 270} = 0.2745\ \text{atm/°R}$$

Therefore, $\Delta H_V = (460 + 280)(0.165 - 0.0429)(0.2745)(°\text{R})$

$$\times\left(\frac{\text{ft}^3}{\text{lb}}\right)\left(\frac{\text{atm}}{°\text{R}}\right).$$

$$= 24.80°\text{R}\left(\frac{\text{ft}^3}{\text{lb}}\right)\left(\frac{\text{atm}}{°\text{R}}\right)\times\left(144\times 14.7\frac{\text{lb}}{\text{ft}^2\cdot\text{atm}}\right)$$

$$= \frac{52{,}496.64}{778}\frac{\frac{\text{ft}\cdot\text{lb}}{\text{lb}}}{\frac{\text{ft}\cdot\text{lb}}{\text{Btu}}} = 67.5\ \text{Btu/lb}$$

1.11 c. Apply Bernoulli's theorem between points A (surface of the liquid) and B (discharge point). Also, take the level at B as the datum plane.

$$\frac{P_A}{\rho_A} + Z_A + \frac{u_A^2}{2g_c} - F + w = \frac{P_B}{\rho_B} + Z_B + \frac{u_B^2}{2g_c}$$

$\dfrac{P_A}{\rho_A} = \dfrac{P_B}{\rho_B} = \dfrac{P_{atm}}{\rho}$ Therefore, these terms cancel out.

$u_A = 0$ (since the cross section of the tank is very large)
$w = 0$ (since no pump is in the line, no work involved)
$Z_B = 0$ (since point B is taken as the datum plane)

Therefore, the Bernoulli equation reduces to $Z_A = F + \dfrac{u_B^2}{2g_c}$.

$$u_B = \sqrt{2g_c(Z_A - 12)} = \sqrt{2\times 32.2\times(20-12)} = 22.7\ \text{ft/s} = 6.92\ \text{m/s}$$

1.12 c. C_{vc} at maximum flow $= 1000\sqrt{\dfrac{1.2}{4}} = 547.7$

C_{vc} at minimum flow $= 125\sqrt{\dfrac{1.2}{30}} = 25.0$

From the list of valves, a 4-in. valve has a maximum-flow coefficient of 775. If this valve is selected,

$$\frac{C_v}{C_{vc}} = \frac{775}{547.7} = 1.415$$

This ratio falls between 1.25 and 2, which is required for good operation of the control valve.

1.13 c. From a table of the properties of steel pipes (Perry or any other source), the i.d. of a Schedule 80, 8-in. commercial steel pipe = 7.625 in. = 7.625/12 = 0.6354 ft.

From table 5-6 (Perry's Handbook),

Absolute roughness of steel pipe = 0.0457 mm
= $0.0457(3.281 \times 10^{-3}) = 0.00015$ ft.

Therefore, the relative roughness is

$$\varepsilon/d_i = \frac{0.00015 \text{ ft}}{0.6354 \text{ ft}} = 0.000236 = 2.36 \times 10^{-4}.$$

1.14 d.

$$d_i = 6.065/12 = 0.5054 \text{ ft}$$

$$\text{Inside cross-sectional area} = \left(\frac{\pi d_i^2}{4}\right) = \frac{\pi(0.5054)^2}{4}$$

$$= 0.2006 \text{ ft}^2$$

$$\text{Density} = \rho = 62.4 \times 0.9 = 56.16 \text{ lb/ft}^3$$

$$\text{Velocity of oil through the pipe} = \frac{500 \text{ gal/min}}{7.48 \text{ gal/ft}^3} \times \frac{1}{60 \text{ s/min}} \times \frac{1}{0.2006 \text{ ft}^2}$$

$$= 5.554 \text{ ft/s}$$

$$\text{Viscosity at flowing temperature} = 4.1\frac{\text{lb}}{\text{h} \cdot \text{ft}} = 4.1\frac{\text{lb}}{\text{h} \cdot 3600\frac{\text{s}}{\text{h}} \cdot \text{ft}}$$

$$= 0.00114\frac{\text{lb}}{\text{s} \cdot \text{ft}}$$

$$\therefore \text{Reynolds number} = \frac{d_i u \rho}{\mu}$$

$$= \frac{0.5052 \times 5.554 \times 56.16}{0.00114} = 138{,}226$$

1.15 b. Apply the Bernoulli equation between the discharge point of the pump and the liquid surface in the tank (see Exhibit 1.15).

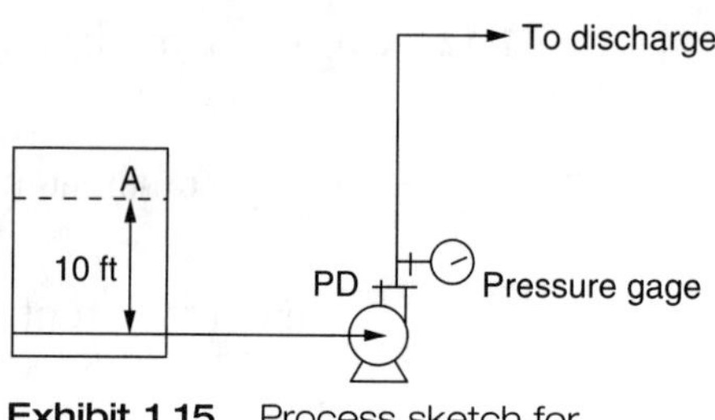

Exhibit 1.15 Process sketch for Problem 1.15

Also assume that the elevation difference between the pump suction point and the discharge point is negligible so that $Z_D = 0$.

$$\frac{P_A}{\rho} + Z_A + \frac{u_A^2}{2g_c} - F + W = \frac{P_D}{\rho} + \frac{u_D^2}{2g_c} + Z_D$$

$$\frac{P_D - P_A}{\rho} = \frac{u_A^2 - u_D^2}{2g_c} - F + W + Z_A - Z_D$$

$$= -\frac{u_D^2}{2g_c} - F + W + Z_A \text{ (since } u_A = 0 \text{ and } Z_D = 0)$$

Velocity through the 2-in. Schedule 40 pipe $= 3(3.068/2.067)^2 = 6.61$ ft/s

$$\therefore \quad \frac{P_D - P_A}{\rho} = -\frac{6.61^2}{2 \times 32.2} - 1.5 + 43.2 + 10 = 51.02 = \text{ft of liquid}$$

Therefore, $P_D - P_A = 51.02 \times \rho = \dfrac{51 \times 100.15}{144} = 35.48 \approx 35.5$ psi.

Then $P_D = 35.5 + P_A = 35.5 + 14.7 = 50.2\,\text{psia} = 35.5\,\text{psig}$. Therefore, the gage shows 35.5 psig pressure.

1.16 b. Assume turbulent flow. Therefore, the orifice coefficient $C_o = 0.61$.

A pitot tube positioned at the center of the pipe's cross section gives the maximum velocity.

$$\text{Density of air} = \frac{29}{359} \times \frac{492}{680} \times \frac{750}{760} = 0.05768 \text{ lb/ft}^3$$

$$\text{4 in. of water column} = \frac{4}{12} \times \frac{62.4}{0.05768} = 360.6 \text{ ft of air}$$

$$u_{max} = 0.98\sqrt{64.4 \times 360.6} = 149.3 \text{ ft/s}$$

$$\text{Average velocity} = u = 0.82(149.3) = 122.4 \text{ ft/s}$$

$$\text{Reynolds number} = \frac{Du\rho}{\mu} = \frac{(20/12)(122.4)(0.05768)}{0.022(0.000672)} = 7.96 \times 10^5$$

Therefore, the flow is turbulent and the calculation of the average velocity is correct.

Flow = 2.182(122.4)(60) = 16,025 cfm (at 750 mm and 220°F)

Therefore, the flow at 60°F and 760 mm pressure

$$= 16{,}025\left(\frac{520}{680}\right)\left(\frac{750}{760}\right) = 12{,}093 \text{ scfm}$$

1.17 c. Discharge side:

$$\text{Discharge-side pressure drop} = \Delta P_\text{d}\left(\sum K + \frac{4fL}{(d_\text{i})_\text{d}}\right)\frac{u_\text{d}^2}{2g_c}$$
$$= \left(8.77 + \frac{4\times 0.0049\times 250}{0.2557}\right)\times\frac{5.21^2}{64.4}$$
$$= 11.77 \text{ ft}$$

$$\text{Suction-side pressure drop} = 0.24\times 2.31 = 0.55 \text{ ft}$$
$$\text{Discharge static head} = 250 \text{ ft}$$

Apply the Bernoulli equation to get the total dynamic head:

$$\text{TDH} = (P_\text{B} - P_1) + \Delta P_\text{s} + \Delta P_\text{d} + Z_\text{B} - Z_\text{A}$$
$$= 0 + 0.55 + 11.77 + 250 - 10$$
$$= 252.32 \text{ ft} = 252.32(0.433) = 109.3 \text{ psi}$$
$$\text{bhp} = \frac{(\text{gpm})(\text{psi})}{1714\times 0.7} = \frac{120\times 109.3}{1714\times 0.7} = 10.93$$

1.18 a. There are three resistances in series.

$$R_1 = \frac{9/12}{0.68(1)} = 1.103 \qquad R_2 = \frac{4/12}{0.15(1)} = 2.222 \qquad R_3 = \frac{8/12}{0.4(1)} = 1.667$$

$$\text{Total resistance} = 1.103 + 2.222 + 1.667 = 4.992$$

The temperature drop is proportional to the resistance.

Resistance of kaolin brick = 1.103
Total temperature drop = 2000 – 180 = 1820°F

Therefore, the temperature drop across the kaolin brick $= \dfrac{1.103}{4.922}\times 1820$

$\approx 408°\text{F}$

Then the interface temperature between the kaolin brick and the insulating brick is

2000 – 408 = 1592°F

1.19 b. The given equation is used to calculate the tube-side convection heat-transfer coefficients for the following conditions:

1. At moderate Δt values
2. For turbulent flow
3. If the fluid flowing through the tube is being cooled

1.20 c. First, calculate the log-mean temperature difference.

ΔT_1 = 450 – 310 = 140°F $\qquad$ ΔT_2 = 350 – 300 = 50°F

$$\Delta T_\text{lm} = \frac{140 - 50}{\ln\frac{140}{50}} = 87.4°\text{F}$$

$$U_{do} = \text{Dirty overall coefficient of heat transfer} = \frac{Q}{A \times \Delta T_{lm}}$$

$$= \frac{490{,}000}{200 \times 87.4} = 28.03 \text{ Btu/h} \cdot \text{ft}^2 \cdot {}^\circ\text{F}$$

1.21 a. Calculate the outlet temperature of the coolant:

Heat duty = 45,000(0.605)(390 – 200) = 5,172,750 Btu/h

Therefore, for the crude stream, $150{,}810(t_o - 100)(0.49) = 5{,}172{,}750$ Btu/h.

$$t_o - 100 = \frac{5{,}172{,}750}{150{,}810(0.49)} = 70^\circ\text{F}$$
$$t_o = 170^\circ\text{F}$$

Calculate the log-mean temperature difference:

Kerosene
390 ⟶ 200
170 ⟵ 100
Crude stream

$$\Delta t_1 = 220 \qquad \Delta t_2 = 100$$
$$\Delta t_{lm} = \frac{220 - 100}{\ln \frac{220}{100}} = 152.2^\circ\text{F}$$

$$R = \frac{390 - 200}{170 - 100} = \frac{190}{70} = 2.71$$
$$S = \frac{170 - 100}{390 - 100} = \frac{70}{290} = 0.241$$

From the correction chart for a one-shell pass and two or more tube passes (from Perry or another source),

$$\text{Correction factor } = F_T \approx 0.905$$
$$\therefore \Delta T_{MTD} = 152.2 \times 0.95 = 137.7^\circ\text{F}$$

1.22 d. i.d. of tube = 0.902 in.
o.d = 1 in.

$$\text{log-mean diameter} = \frac{d_o - d_i}{\ln \frac{d_o}{d_i}} = \frac{1 - 0.902}{\ln \frac{1}{0.902}} = 0.9502 \text{ in.}$$

The dirty overall heat transfer coefficient is given by

$$\frac{1}{U_{\text{do}}} = \frac{1}{2000} + \frac{1}{2000} + \frac{0.049/12}{63} \times \frac{1}{0.9502} + \frac{1}{2000} \times \frac{1}{0.902} + \frac{1}{1800} \times \frac{1}{0.902}$$
$$= 0.0005 + 0.0005 + 0.0000682 + 0.000554 + 0.000616$$
$$= 0.002238$$

Therefore, dirty U_{do} = 446.8 Btu/h·ft^2·°F.

When the tubes are cleaned on both sides, dirt resistances disappear, and then

$$\frac{1}{U_{\text{c}}} = \frac{1}{2000} + \frac{0.049/12}{63} \times \frac{1}{0.9502} + \frac{1}{1800} \times \frac{1}{0.902}$$
$$= 0.0005 + 0.000682 + 0.000616$$
$$= 0.001798$$
$$U_{\text{c}} = 556.2 \text{ Btu/h·ft}^2\text{·°F}$$

Therefore, after thoroughly cleaning both sides,

$$\text{Percent increase in overall coefficient} = \frac{556.2 - 446.8}{446.8} \times 100 = 24.5\%$$

1.23 b.

$$d_{\text{o}} = 115 \text{ mm} = 0.115 \text{ m}$$
$$T_{\text{surroundings}} = 273.1 + 21.1 = 294.2 \text{ K}$$
$$\Delta t = 400 - 294.2 = 105.8 \text{ K}$$

Calculate h_{c}:

$$h_{\text{c}} = 1.0813\left(\frac{105.8}{0.115}\right)^{0.25} = 5.51\frac{\text{W}}{\text{m}^2 \cdot \text{K}}$$

Calculate h_{r}:

$$h_{\text{r}} = \sigma\varepsilon\left(T_{\text{s}}^4 - T_{\text{a}}^4\right)\big/(T_{\text{s}} - T_{\text{a}})$$
$$= 5.67 \times 0.8(4^4 - 2.942^4)/(400 - 294.2)$$
$$= 7.76\frac{\text{W}}{\text{m}^2 \cdot \text{K}}$$

Therefore, $h_{\text{a}} = h_{\text{c}} + h_{\text{r}}$ = 5.51 + 7.76 = 13.27 W/m^2·K.

Area per meter of tube = $\pi d_{\text{o}} L = \pi(0.115)(1)$ = 0.3613 m^2

Heat loss per meter of tube = $h_a A \Delta t$

= 13.27 × 0.3613 × 105.8 = 507.3 W/m

1.24 a. For the diffusion of one component with the second component non-diffusing, the rate of diffusion of the diffusing component (in this example, oxygen) is given by

$$N_A = \frac{D_{AB}P}{RTzp_{BM}}(p_{A1} - p_{A2})$$

In this example, $P = 1$ atm and $T = 460 + 32°R$.

$$p_{A1} = 0.13 \text{ atm} \quad p_{B1} = 1 - 0.13 = 0.87 \text{ atm}$$
$$p_{A2} = 0.065 \text{ atm} \quad p_{B2} = 1 - 0.065 = 0.935 \text{ atm}$$
$$p_{BM} = \frac{p_{B2} - p_{B1}}{\ln\left(\frac{p_{B2}}{p_{B1}}\right)} = \frac{0.935 - 0.87}{\ln\,(0.935/0.87)} = 0.902 \text{ atm}$$

Rate of diffusion of oxygen $= N_A$

$$= \frac{0.7224(1)}{0.7302(492)(0.12/12)(0.902)}(0.13 - 0.065)$$
$$= 0.0145 \text{ (lb mol)/(h} \cdot \text{ft}^2)$$

1.25 b. Mole fraction of water in the liquid mixture $= \dfrac{95/18.02}{95/18.02 + 5/71.14} = 0.987$

From the t-x diagram, the bubble point is found to be about 89.8 to 90°C.

1.26 a.

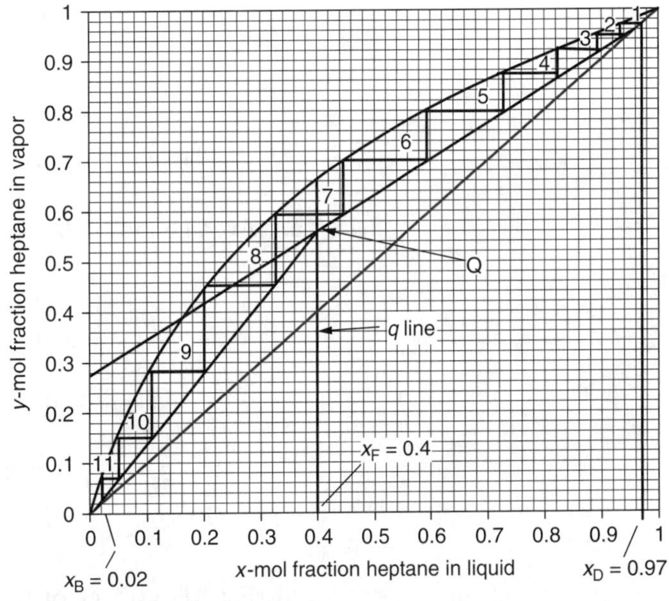

Exhibit 1.26b Construction of the theoretical stages, Problem 1.26

For a reflux ratio of 2.5, the intercept on the y-axis is

$$\frac{x_D}{R+1} = \frac{0.97}{2.5+1} = 0.277$$

A line is drawn from $x_D = 0.97 = y_D$ to intersect with the y-axis at $y = 0.277$. This is the operating line for the rectifying section. It meets the q line at point Q.

A line is drawn from $x_B = y_B = 0.02$ to meet the q line and the rectifying-section operating line at point Q. This is the operating line for the stripping section. The theoretical stages are then stepped down, as in Exhibit 1.26b.

From the stage construction, the number of stages = 11.

1.27 b. Calculation of N_{OG}:

$$y_1 = 0.1 \quad y_2 = 0.00111 \qquad y_1^* = 1.406(0.04523) = 0.0636 \qquad y_2^* = 0$$

$$N_{OG} = \frac{y_1 - y_2}{(y - y^*)_{lm}}$$

$$(y - y^*)_{lm} = \frac{(0.1 - 0.0636) - (0.0011 - 0)}{\ln \frac{0.1 - 0.0636}{0.00111 - 0}} = 0.01011$$

$$\frac{y_1 - y_2}{(y - y^*)_{lm}} = \frac{0.1 - 0.00111}{0.01011} = 9.78$$

$$G = 950.5/23.76 = 40 \text{ (lb mol)/(h·ft}^2)$$

$$1 - y_1 = 0.9 \qquad 1 - y_2 = 0.9989$$

$$(1 - y)_{lm} = \frac{0.9989 - 0.9}{\ln \frac{0.9989}{0.9}} = 0.9486$$

$$H_{OG} = \frac{G}{K_y a(1 - y_{lm})} = \frac{40}{16.2(0.9486)} = 2.6$$

Height $= Z = H_{OG}(N_{OG}) = 9.78(2.6) = 25.43$ ft ≈ 25.5 ft

1.28 a. Average molecular weight = 0.05(17) + 29(0.95) = 28.4

$$\rho_g = \left(\frac{28.4}{359}\right)\left(\frac{492}{528}\right)\left(\frac{15}{14.7}\right) = 0.0752 \text{ lb/ft}^3$$

Gas flow rate = 30,000(0.0752) = 2256 lb/h

$$\frac{L}{G}\sqrt{\frac{\rho_g}{\rho_L}} = 1\sqrt{\frac{0.0752}{62.4}} = 0.0347 \qquad \frac{\rho_w}{\rho_L} = \frac{62.4}{50} = 1.248$$

For $\Delta P = 0.5$ in. of H_2O, Y (ordinate) = 0.058 (from the pressure drop chart).

$$G_L^2 = \frac{0.0752(50)(4.18 \times 10^8)(0.058)}{52 \times 1.248 \times 0.23^2} = 1{,}884{,}640 \qquad G_L = 1373 \text{ lb/h} \cdot \text{ft}^2$$

$$A = \frac{2256}{1373} = 1.643 \text{ ft}^2$$

$$D = \sqrt{\frac{1.643}{0.7854}} = 1.446 \approx 1.5 \text{ ft} = 18 \text{ in.}$$

1.29 b. The solution of the rate equation for a unimolecular reversible reaction is as follows:

$$-\ln\left(1 - \frac{X_A}{X_{AE}}\right) = \frac{M}{M + X_{AE}} k_1 t$$

where

X_A = conversion at time t

X_{AE} = equilibrium conversion

$M = C_{B0}/C_{A0}$

A plot of $-\ln\left(1-\frac{X_A}{X_{AE}}\right)$ versus t should be a straight line whose slope is

$$\text{slope} = \frac{M}{M+X_{AE}}k_1$$

The given plot of the data shows a straight line.

Slope of the line = (5.2 – 0)/(4.2 – 0) = 1.2381

$M = 0.5/0.8 = 0.625$

Therefore, $$\frac{M+1}{M+X_{AE}}k_1 = 1.2381$$

Then $$\frac{0.625+1}{0.625+0.35}k_1 = 1.2381$$

$$k_1 = \frac{1.2381}{1.667} = 0.7429\ \text{h}^{-1}$$

1.30 a. Arrhenius relation: $k = \alpha e^{-E/RT}$

Taking the logarithm of both sides yields $\ln k = \ln\alpha - \frac{E}{R}\left(\frac{1}{T}\right)$.

A plot of ln k versus $1/T$ is already given.

The slope of the line is $-\frac{E}{R} = -6.579$.

Therefore, $E = 6.579 \times (1 \times 10^3) \times 1.987 = 13{,}072$ cal/g mol

1.31 b. The rate of the reaction, in terms of the conversion, can be written as

$$-r_A = kC_A^2 = C_{A0}\frac{dX_A}{dt} = C_{A0}^2(1-X_A)^2$$

Therefore, $$\frac{dX_A}{dt} = kC_{A0}(1-X_A)^2.$$

For a mixed reactor,

$$\tau = \frac{V}{v_0} = \frac{C_{A0}(1-X_{Af})}{kC_{A0}(1-X_{Af})^2} = \frac{1}{k(1-X_{Af})}$$

where

X_{Af} = concentration of effluent from the mixed reactor

Now write a similar equation for the large reactor and take the ratio of the two to obtain the following:

$$\frac{\tau_6}{\tau_1}=\frac{V_2}{V_1}=\frac{\frac{1}{1-X_{Af2}}}{\frac{1}{1-X_{Af1}}}=\frac{1-X_{Af1}}{1-X_{Af2}}=\frac{6}{1}$$

$$1-X_{Af2}=(1-X_{Af1})/6=\frac{1-0.5}{6}=0.5/6$$

$$X_{Af2}=1-0.5/6=0.917$$

$$\%\ \text{conversion}=0.917\times 100=91.7\%$$

1.32 a. For a plug-flow reactor, space-time is given by

$$\tau=C_{A0}\int_0^{X_A}\frac{dX_A}{kC_{A0}(1-X_A)^2}=\frac{1}{k}\left[\frac{X_A}{1-X_A}\right]$$

Since everything else is the same, taking the ratio again yields

$$\frac{\tau_P}{\tau_m}=\frac{\frac{1}{k}\frac{X_A}{1-X_A}}{\frac{1}{k}\frac{1}{1-X_{Af1}}}=1$$

since τ is the same for both mixed- and plug-flow reactors.

$$\frac{X_{Ap}}{1-X_{Ap}}=\frac{1}{1-X_{Af1}}=\frac{1}{1-0.5}=2$$

From which

$$X_{Ap}=0.67.$$

$$\text{Percent conversion}=0.67(100)=67\%$$

1.33 a. For reactions taking place in parallel, the concentration levels of the reactants are the key to properly control the product distribution. Divide the second equation by the first equation to obtain

$$\frac{r_S}{r_R}=\frac{k_2}{k_1}C_A^{-0.5}C_B^{1.5}$$

This ratio has to be as low as possible to favor the first reaction. For this to happen, C_A should be high and C_B should be low. Also, the concentration dependency of B is more pronounced than that of A, so it is more important to have less B than to have high C_A.

Therefore, method a (C_A high, C_B low) should be used.

1.34 b. Total depreciation = S = 50,000 – 8000 = \$42,000
Yearly payout = R
Number of payments = 10
Annual interest rate = i = 8.25% = 0.0825

$$R=S\frac{i}{(1+i)^n-1}=42{,}000\times\frac{0.0825}{(1+0.0825)^{10}-1}=\$2865$$

1.35 c. Uniform corrosion, due to chemical or electrochemical attack, takes place uniformly on the surface of the material. When the corrosion rate is known, the equipment life is predictable. This type of corrosion can be prevented by choosing the proper construction material and using inhibitors or cathodic protection. Alternatively, a corrosion allowance can be added to the calculated thickness to get the final design plate thickness.

1.36 b. When a step change is applied to the system, the transform of $y(t)$ becomes

$$Y(s) = \frac{1}{s}\left(\frac{\tau_1 s + 1}{\tau_2 s + 1}\right)$$

By the final-value theorem, $\lim_{t\to 0}[y(t)] = \lim_{s\to\infty}[sf(s)]$.

$$= \lim_{s\to 0} s\left[\frac{1}{s}\left(\frac{\tau_1 s + 1}{\tau_2 s + 1}\right)\right]$$

$$= 1$$

By the initial-value theorem, $\lim_{t\to 0}[y(t)] = \lim_{s\to\infty}[sf(s)]$.

$$= \lim_{s\to\infty} s\left[\frac{1}{s}\left(\frac{\tau_1 s + 1}{\tau_2 s + 1}\right)\right]$$

$$= \lim_{s\to\infty} \frac{\tau_1 + 1/s}{\tau_2 + 1/s}$$

$$= \frac{\tau_1}{\tau_2} = 5 \text{ (given)}$$

The value of $y(t)$ varies between 5 and 1. The maximum value is therefore 5.

1.37 b. First, calculate the LFL of the mixture. For this calculation, percentage compositions of the combustibles on an air-free basis are required.

Total combustibles = 2 + 0.5 + 1 = 3.5%
mol % methane = (2/3.5)(100) = 57.14%
mol % ethylene = (0.5/3.5)(100) = 14.29%
mol % hexane = (1/3.5)(100) = 28.57%

$$\text{LFL}_{\text{mixture}} = \frac{100}{(57.14)/5 + (14.29)/2.7 + (28.57)/1.1} = 2.34\%$$

$$\text{MOC} = \text{LFL}\left(\frac{\text{stoichiometric mol of oxygen}}{1 \text{ mol of fuel}}\right)$$

Calculation of oxygen per mole of fuel:

$CH_4 + 2O_2 \rightarrow CO_2 + 2H_2O$ 2 mol of O_2
$C_2H_4 + 3O_2 \rightarrow 2CO_2 + 2H_2O$ 3 mol of O_2
$C_6H_{14} + 9.5O_2 \rightarrow 6CO_2 + 7H_2O$ 9.5 mol of O_2

O_2 per mole of fuel = 0.5714(2) + 0.1429(3) + 0.2857(9.5)
= 4.286 mol/(mol fuel)

Therefore,

Minimum O_2 concentration (MOC) for combustion = 2.34(4.286)
= 10.03%

1.38 **b.** PSV loading for a blocked discharge = Pump design capacity (rated flow)
= 743 gpm

1.39 **c.** For steam, when k_{sh} (superheated steam correction factor) = 1 and k_b (back pressure factor) = 1 (given), the orifice area is then given by

$$A = \frac{W_S}{51.5K_D P} \quad \text{or} \quad W_S = 51.5K_D PA$$

Therefore, 1000 = $51.5K_DPA$, and then K_DPA = 1000/51.5 = 19.42.

For NH_3, the relieving rate $= C\ K_D PA\sqrt{\frac{M}{T}}$ lb/h

M = 17.03 for nitrogen
T = 460 +150 = 610°R

$$C = 520\sqrt{k\left(\frac{2}{k+1}\right)^{\frac{k+1}{k-1}}} = 520\sqrt{1.33\left(\frac{2}{1.33+1}\right)^{\frac{1.33+1}{1.33-1}}} \approx 350$$

Then, the relieving capacity for ammonia = $350(19.42)\sqrt{\frac{17.03}{610}}$ = 1135.7 lb/h.

1.40 **a.** The maximum allowable working pressure (MAWP) should be calculated on the basis of the corroded thickness.

The corroded thickness is

$$t_c = t - \text{corrosion allowance} = 0.625 - 0.125 = 0.5 \text{ in.}$$

$$\text{MAWP} = P = \frac{SEt}{R + 0.6t} = \frac{13{,}800 \times 0.8 \times 0.5}{48 + 0.6 \times 0.5} = 114.3 \text{ psig}$$

(Please note that the calculated P is, in reality, the difference between the internal pressure and the external pressure. The external pressure is usually 1 atm, or 14.7 psia, and hence the calculated internal pressure differential becomes the internal pressure in psig.)

AFTERNOON EXAM (CHAPTER 2) SOLUTIONS

2.1 **b.** First, calculate the amount of flue gas and its composition. For this, prepare a table as follows:

Basis: 100 lb mol of fuel gas

Component	Feed, lb mol	CO_2, lb mol	H_2O, lb mol	O_2, lb mol	N_2, lb mol
CH_4	97.0	97.0	194	194	
O_2	1.6			–1.6	
N_2	1.4				3.6
Total	100.0	97.00	194.0	192.4	3.6

$$\text{Total oxygen used} = 1.3(192.4) = 250.12 \text{ lb mol}$$
$$\text{Excess oxygen} = 0.3(250.12) = 75.04 \text{ lb mol}$$
$$\text{Nitrogen in air} = 3.76(250.12) = 940.45 \text{ lb mol}$$
$$\text{Total nitrogen in flue gas} = 940.45 + 3.6 = 944.05 \text{ lb mol}$$

$$\text{Flue gas} = 97 + 194 + 75.04 + 944.05 = 1310.1 \text{ lb mol}$$
$$\text{Partial pressure of water} = (194/1310.1) \times 14.7 \approx 2.18 \text{ psia}$$

From the steam tables, at this pressure, the temperature by interpolation is

Temperature t, °F	Vapor Pressure p, psia
120	1.6945
130	2.225

At a vapor pressure of 2.18 psia,

$$t = \frac{2.18 - 1.6945}{2.225 - 1.6945} \times 10 + 120 = 9.15 + 120 \approx 129.2°\text{F}$$

At this temperature, the water will start condensing. Hence, the dew point temperature = 129.2°F.

2.2 **c.** $Al_2O_3 + 3H_2SO_4 \rightarrow Al_2(SO_4)_3 + 3H_2O$
101.96 3 × 98.08 342.14 3 × 18.03

$$Al_2O_3 \text{ in bauxite feed} = 0.554(2160) = 1196.64 \text{ lb}$$
$$Al_2(SO_4)_3 \text{ produced} = 3600 \text{ lb}$$
$$Al_2O_3 \text{ used} = \frac{101.96}{342.14} \times 3600 = 1072.8 \text{ lb}$$
$$\% \; Al_2O_3 \text{ used} = \frac{1072.8}{1196.64} \times 100 = 89.65\% \approx 89.7\%$$

2.3 **b.** Basis: 100 mol of fresh feed

Overall balance: $100 = M + P$

where
M = moles of NH_3 produced, and P = purge, mol/h

Overall H_2 balance: $100(0.751) = M(0) + P(0.8)$

From which

$$P = \frac{100(0.751)}{0.8} = 93.875 \text{ mol/h}$$

Then $M = 100 - 93.875 = 6.125$ mol/h

Now take the hydrogen balance around the mixing point of the fresh feed and recycle:

$$100(0.751) + R(0.800) = (100 + R)(0.795)$$
$$75.1 + R(0.800) = (100 + R)(0.795)$$

From which,

$$R(0.800 - 0.795) = 79.5 - 75.1 = 4.4$$
$$R = 4.4/0.005 = 880 \text{ mol/h.}$$

2.4 **a.** Draw a sketch similar to the one in Exhibit 2.4. Take the material balance around the nth plate and the product, as shown in the sketch. The following equations can be written:

Overall balance: $V_{n+1} = L_n + D$

Component A balance: $V_{n+1}y_{n+1} = L_n x_n + Dx_D$

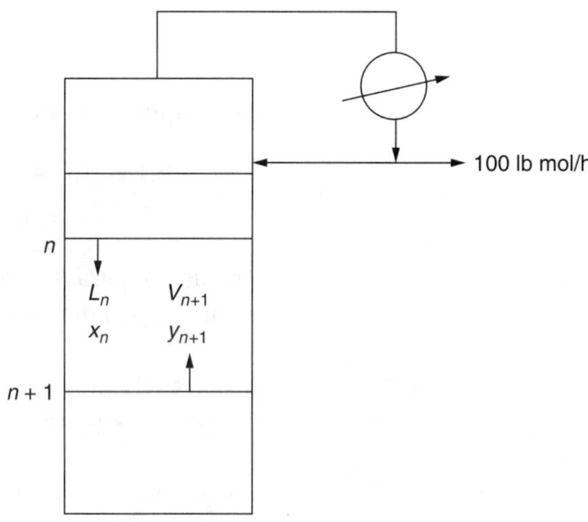

Exhibit 2.4 Material balance for rectification section of distillation column for Problem 2.4

Substitute the available data and rewrite the equations as

$$V_{n+1} = L_n + 100$$
$$V_{n+1} = L_n + 100V_{n+1}(0.7514) = L_n(0.66) + D(0.98)$$

Dividing the second equation by 0.7514,

$$V_{n+1} = L_n(0.87836) + D(130.423)$$

Solving these two equations for L_n,

$$L_n = \frac{-30.423}{-0.12164} \approx 250 \text{ lb mol}$$

Therefore, the reflux ratio used = 250/100 = 2.5.

2.5 **b.** The van der Waals equation is

$$\left(P+\frac{n^2a}{V}\right)(V-nb)=nRT$$

$$P=14.7+5.3=20 \text{ psia}$$

Substituting the given values into the equation yields

$$\left[20+\frac{0.0044^2\times(5.1\times10^3)}{1.82}\right](1.82-0.0044\times0.516)$$
$$=0.0044(10.73)T$$

From which, T = 772°R = 772 – 460 = 312°F.

2.6 **b.** Assume that 77°F is the datum temperature.

H of $CaCO_3$ at 77°F = 0
H of $CaCO_3$ at 400°F = (400 – 77) × 0.242 = 78.17 Btu/lb
H of CO_2 at 1832°F = (1832 – 77)(0.2830) = 496.7 Btu/lb
H of CaO at 1832°F = (1832 – 77)(0.245) = 430 Btu/lb

By energy balance,

$$\begin{aligned} Q \text{ to be supplied} &= \text{Heat of reaction} \\ &\quad + \text{Enthalpy of products} - \text{Enthalpy of reactants} \\ &= 20{,}000(788.4)+8800(496.7)+11{,}200(430.7) \\ &\quad -20{,}000(78.17) \\ &= 23.4\times10^6 \text{ Btu/h} \end{aligned}$$

With a kiln efficiency of 80%,

Heat to be supplied = $(23.4 \times 10^6)/0.8 = 29.25 \times 10^6$ Btu/h
= 29.25 MMBtu/h ≈ 29.3 MMBtu/h

2.7 **c.** By definition, the heating value of a fuel is the negative of the heat of combustion. The lower heating value (LHV) is a value based on water vapor as the product. When the product water is liquid at 25°C and 1 atm, the higher heating value (HHV) is obtained.

Basis of calculation: 1 g mol of total fuel gas.

LHV of methane = 802 kJ/g mol
LHV of ethane = 1428 kJ/g mol
λ of water at 25°C and 1 atm = 44.013 kJ/kg mol

HHV of CH_4 = 802 + 2(44.013) = 890.03 kJ/g mol
HHV of C_2H_6 = 1428 + 3(44.013) = 1560.04 kJ/g mol

N_2 has no heating value.

$$\text{HHV/g mol of total fuel gas} = 0.85(890.03) + 0.12(1560.04)$$
$$= 943.73 \text{ kJ/g mol} = 943.73 \times 10^3 \text{ kJ/kg mol}$$

$$\text{HHV/m}^3 \text{ of fuel gas at standard conditions} = \frac{943.73 \times 10^3 \frac{\text{kJ}}{\text{kg·mol}}}{22.414 \frac{\text{m}^3}{\text{kg·mol}}}$$
$$= 42{,}100 \text{ kJ/m}^3$$

2.8 d. The heat to be supplied to heat the air = 70,000(0.24)(150 – 80)
= 1,176,000 Btu/h

From the steam tables, the enthalpies of steam and steam condensate are

H_s = 1180.2 Btu/lb of saturated steam at 300°F
h_{cond} = 249.18 Btu/lb at 280°F

Therefore, the heat available to heat the air/lb of steam = 1180.2 – 249.18 = 931.02 Btu/lb.

$$\text{Steam condensed} = \frac{1{,}176{,}000}{931.02} = 1263 \text{ lb/h}$$

2.9 b. $T_{br} = \dfrac{486.6}{699} = 0.6961 \qquad P_r = \dfrac{1}{31.4} = 0.032$

From the Z chart, $Z = 0.97$.

Substituting the appropriate values into the reduced equation,

$$\frac{\Delta H_V}{ZRT_C} = \frac{B}{T_C}\left[\frac{T_r}{T_r + C/T_C}\right]^2$$

$$\Delta H_V = ZRB\left[\frac{T_r}{T_r + C/T_C}\right]^2$$

$$= (0.97)(8.314)(4104.84)\left[\frac{0.6961}{0.6961 + \left(\frac{-74.56}{699}\right)}\right]^2$$

$$= \frac{46{,}170 \text{ J}}{1054 \text{ J/Btu}} \times \frac{454.6 \text{ g/lb}}{\text{g mol}}$$

$$= 19{,}906 \text{ Btu/lb mol} = 132.6 \text{ Btu/lb}$$

2.10 a. The reactions and their heats of reaction are listed in the following table:

	$\Delta H_r = \Sigma(\Delta H_f)_p - \Sigma(\Delta H_f)_R$	Heats of Reactions (at 25°C and 1 atm), kcal/g mol
(1)	$CH_3CHO(g) + H_2(g) \rightarrow C_2H_5OH(g)$	$\Delta H_1 = -12.51$
(2)	$C_2H_5OH(g) + 3O_2(g) \rightarrow 2CO_2(g) + H_2O(l)$	$\Delta H_2 = -340.83$
(3)	$H_2(g) + \frac{1}{2}O_2(g) \rightarrow H_2O(l)$	$\Delta H_3 = -68.32$
(4)	$H_2O(l) \rightarrow H_2O(g)$	$\Delta H_4 = +10.52$

Add the first two equations and subtract the third to obtain the fifth equation:

(1)	$CH_3CHO(g) + H_2(g) \rightarrow C_2H_5OH(g)$	$\Delta H_1 = -12.51$
(2)	$C_2H_5OH(g) + 3O_2(g) \rightarrow 2CO_2(g) + 3H_2O(l)$	$\Delta H_2 = -340.83$
(3)	$-[H_2(g) + \frac{1}{2}O_2(g) \rightarrow H_2O(l)]$	$-[\Delta H_3 = -68.32]$
	$= (1) + (2) - (3)$	
(5)	$CH_3CHO(g) + 2\frac{1}{2}O_2 \rightarrow 2CO_2(g) + 2H_2O(l)$	$\Delta H_c = \Delta H_1 + \Delta H_2 - \Delta H_3$ $= -12.51 - 340.83 + 68.32$ $= -285.02$ kcal/g mol

2.11 c. $$\text{Velocity through pipe} = u = \frac{125\text{ m}^3/\text{h}}{(3600\text{ s/h})(0.0176715\text{ m}^2)} = 1.965\text{ m/s}$$

$$\text{Reynolds number} = \frac{Du\rho}{\mu} = \frac{0.15(1.965)(1250)}{0.0012} = 3.07\times10^5$$

$$\left[\textit{Note}:\ 1\text{ cP} = 0.01\text{ poise} = \frac{0.01\text{g}}{1\text{cm}\cdot\text{s}} = \frac{(0.01\text{g})(1\times10^{-3}\text{kg/g})}{(1\text{cm})\left(1\times10^{-2}\frac{\text{m}}{\text{cm}}\cdot\text{s}\right)} = 1\times10^{-3}\frac{\text{kg}}{\text{m}\cdot\text{s}}\right]$$

Relative roughness of the steel pipe,

$$\frac{\varepsilon}{D} = \frac{(0.0457\text{ mm})(1\times10^{-3}\text{m/mm})}{(0.15\text{m})} = 0.000305$$

From the Fanning friction factor chart, $f = 0.0043 = 4.3 \times 10^{-3}$.

2.12 a. At Reynolds number $= 2\times10^5$ and relative roughness $= 0.00045$, Fanning friction factor, $f = 0.0046$ (from the Fanning friction chart).

$$\text{Reynolds number} = \frac{Du\rho}{\mu} = 2.6\times10^5$$

$$u = \frac{2.6\times10^5(0.65\times0.000672)}{0.3355(56.2)} = 6.023\text{ ft/s}$$

$$\Delta P = \frac{2fLu^2\rho}{144g_cD} = \frac{2(0.0046)(100)(6.023^2)(56.2)}{144(32.2)(0.3355)} = (1.206\text{ psi})/(100\text{ ft})$$

2.13 d. The design pressure, and not the operating pressure, should be used to determine the maximum pressure. Also, HLL is to be used to calculate the suction head.

Design pressure = 50 psig = 64.7 psia
Suction head = (3 + 18)(0.433)(0.8) ≈ 7.3 psi
Line loss (taking the minimum, 1.5 psi) = 1.5 psi

Therefore, the maximum suction pressure = 64.7 + 7.3 – 1.5
= 70.5 psia
= 55.8 psig ≈ 56 psig

2.14 b. At the intersection of the *H-Q* curve and the system curve, or the operating point,

Flow rate ≈ 70 gpm
Head ≈ 86 ft

The corresponding efficiency read from the efficiency curve is 42%. Then,

$$\Delta P = 86 \times 0.433 = 37.24 \text{ psi}$$

$$\text{bhp} = \frac{70 \times 37.24}{1714(0.42)} = 3.62$$

With 80% motor efficiency, the motor size needed is

$$\text{mhp} = \frac{3.62}{0.8} = 4.53$$

The next standard size motor is 5 hp. Therefore, a 5-hp motor is recommended.

2.15 b. The i.d. of a 4-in. Schedule 40 pipe is

(4.026 in.) × (2.54 cm/in.) = 10.226 cm = 0.10226 m

$$\beta = \frac{5}{10.226} = 0.489 \qquad \beta^4 = 0.05716 \qquad 1 - \beta^4 = 0.9428$$

$$h = \frac{18.5 - 4}{20 - 4} \times 250 = 226.6 \text{ cm} = 2.266 \text{ m}$$

Assume that the flow is fully turbulent, which means that the orifice coefficient $C_o = 0.61$.

$$u_o = 0.61\sqrt{\frac{2 \times g \times \Delta H}{1 - \beta^4}} = 0.61\sqrt{\frac{2 \times 9.81 \times 2.266}{0.9428}} = 4.2 \text{ m/s}$$

Check the Reynolds number.

$$\frac{Du\rho}{\mu} = \frac{0.05 \times 4.2 \times 1000}{1 \times 10^{-3}} = 3.1 \times 10^6$$

Therefore, the flow is turbulent and $C_o = 0.61$.

Flow = $0.7854(0.05)^2(4.2)(3600) = 29.7 \text{ m}^3\text{/h}$

2.16 c. At maximum flow, ΔP_c is minimized and $\Delta P_c = 50$ psi.
At minimum flow, ΔP_c is maximized and $\Delta P_c = 110$ psi.

$$C_{vc} \text{ at maximum flow} = 1200\sqrt{\frac{0.9}{50}} = 161$$

$$C_{vc} \text{ at minimum flow} = 100\sqrt{\frac{0.9}{110}} = 9.0$$

$$\text{Required rangeability} = 161/9 = 17.9 \approx 18$$

2.17 b. $\text{Flow area of channel} = 2 \times 2 - 0.7854 \times 0.84^2$
$= 3.446 \text{ in.}^2 = 0.02393 \text{ ft}^2$

$$\text{Equivalent diameter} = 4r_H = \frac{4 \times 3.446}{[8 + \pi(0.84)]}$$
$$= 1.2956 \text{ in.} \approx 1.3 \text{ in} = 0.108 \text{ ft}$$

$$u = \frac{100}{60 \times 7.48 \times 0.02393} = 9.31 \text{ ft/s}$$

$$\frac{Du\rho}{\mu} = \frac{0.108 \times 9.31 \times 62.4}{0.9 \times 0.000672} = 1.04 \times 10^5$$

$f' = 0.0236$ (from the Moody chart at a Reynolds number of 1.04×105)

$$\Delta P = \frac{0.0236 \times 50 \times 9.31^2 \times 62.4}{64.4 \times 0.108 \times 144} = 6.34 \text{ psi}$$

2.18 d. The Prandtl number is a dimensionless quantity, and consistent units must be used to calculate it. Consistent units can be obtained in the following manner:

$$C_P = 0.615 \text{ Btu/lb} \cdot {}^\circ\text{F}$$

$$k = (1.55 \times 10^{-6}) \frac{\text{Btu}}{\left(\frac{\text{s}}{3600 \text{ s/h}}\right)\left(\frac{\text{in.}}{12 \text{ in./ft}}\right)({}^\circ\text{F})}$$

$$= 0.067 \frac{\text{Btu}}{\text{h} \cdot \text{ft} \cdot {}^\circ\text{F}} = 0.067 \frac{\text{Btu}}{\text{h} \cdot \text{ft}^2 \cdot ({}^\circ\text{F/ft})}$$

$$\mu = 3.05 \times 2.42 = 7.38 \frac{\text{lb}}{\text{h} \cdot \text{ft}}$$

$$\text{Prandtl number} = \frac{C_P\mu}{k} = \frac{0.615 \times 7.38}{0.067} = 67.74 \text{ (dimensionless)}$$

2.19 b. Inside surface area of the sphere $= \pi D_i^2 = \pi(4/12)^2 = 0.3461\ \text{ft}^2$

Outside surface area of the sphere $= \pi D_o^2 = \pi(8/12)^2 = 1.4103\ \text{ft}^2$

$$\text{Mean area } A_m = \sqrt{A_i A_o} = 0.6986\ \text{ft}^2$$

$$\text{Heat loss} = Q = \frac{k_m A_m \Delta t}{\Delta X} = \frac{26(0.6986)(300 - 220)}{2/12} = 8719\ \text{Btu/h}$$

2.20 b. If the steam film and the wall resistances are assumed negligible, the pipewall surface temperature is 366°F.

$$\text{Surface temperature} = 460 + 366 = 826°\text{R}$$
$$\text{Surrounding air temperature} = 460 + 70 = 530°\text{R}$$
$$\Delta t = 366 - 70 = 296°\text{F}$$

$$\text{Convection heat-transfer coefficient} = 0.5\left(\frac{\Delta t}{d_o}\right)^{0.25} = 0.5\left(\frac{296}{2.375}\right)^{0.25} = 1.67\ \text{Btu/(h}\cdot\text{ft}^2\cdot°\text{F)}$$

$$\text{Radiation heat-transfer coefficient} = 0.8 \times 0.1713(8.26^4 - 5.3^4)/(296) = 1.79\ \text{Btu/(h}\cdot\text{ft}^2\cdot°\text{F)}$$

Therefore, the surface coefficient of heat transfer,

$$h_a = h_c + h_r = 1.67 + 1.79 = 3.46\ \text{Btu/(h}\cdot\text{ft}^2\cdot°\text{F)}$$

$$\text{Surface area per foot of pipe} = \pi d_o L = \pi \times \frac{2.375}{12} \times 1 = 0.6218\ \text{ft}^2 \text{ per foot of pipe}$$

$$\text{Heat loss per foot of pipe} = 3.46\,(0.6218)(296) = 636.8\ \text{Btu/h per foot of pipe}$$

2.21 b. The expression for the Grashof number is given by

$$N_{Gr} = \frac{D_o^3 \rho_f^2 g \beta_f \Delta t}{\mu_f^2}$$

$t_f = (220 + 100)/2 = 160°\text{F}$ $\quad \Delta t = 220 - 100 = 120°\text{F}$

$\rho_f = 61.28\ \text{lb/ft}^3$ $\quad \mu_f = 1.221\ \text{cP}$ $\quad g = 4.18 \times 10^8\ \text{lb/h}\cdot\text{ft}$

$$\beta_f = \frac{0.0164 - 0.01624}{\frac{0.0164 + 0.01624}{2}(20)} = 6.135 \times 10^{-4}\ \frac{1}{°\text{F}}$$

$$N_{Gr} = \frac{0.0833^3 \times 61.28^2 \times 4.18 \times 10^8 \times 6.135 \times 10^{-4} \times 120}{(1.221 \times 2.42)^2} = 7.65 \times 10^6$$

2.22 a. The dew point of the mixture from the given condensation data is 213.8°F. Similarly, the bubble point is 185°F. The condensing range is therefore 213.8 to 185°F.

Heat load in the condensing range = 13.9792 – 1.2874
= 12.1518 MMBtu/h

2.23 b. Since the steam film resistance and the wall resistance are negligible,

$$U = h_i$$

When all the conditions are the same except the flow rate,

$$U = h_i \propto G^{0.8} \text{ (turbulent flow)}$$

Therefore $$\frac{U_2}{U_1} = \left(\frac{G_2}{G_1}\right)^{0.8} = 1.5^{0.8} = 1.3832 \text{ and } W_2 = 1.5W_1$$

where 1 and 2 refer to the two cases.

Then the following can be written:

$$\frac{\ln\frac{261-80}{261-t_2'}}{\ln\frac{261-80}{261-170}} = \frac{\frac{U_2A}{1.5WC_P}}{\frac{U_1A}{WC_P}} = 0.9221$$

Simplifying, $$\frac{\ln\frac{261-80}{261-t_2'}}{0.68764} = \frac{\frac{U_2A}{1.5WC_P}}{\frac{U_1A}{WC_P}} = 0.9221$$

Then, $$\ln\frac{261-80}{261-t_2'} = 0.9221(0.68764) = 0.6341$$

$$\frac{261-80}{261-t_2'} = e^{0.6341} = 1.8853$$

And $$181 = 492.1 - 1.8853\, t_2'$$

Therefore, $$t_2' = \frac{492.1-181}{1.8853} = 165°\text{F}$$

where t_2' is the outlet temperature of the cold liquid when the flow is increased by 50%.

2.24 d. The relationship between k_G' and k_G is given by the following relation:

$$k_G'P = k_G p_{BM}$$

where P = 1 atm.

$$p_{B1} = 1 - 0.2 = 0.8$$

$$p_{B2} = 1 - 0.03 = 0.97$$

$$p_{BM} = \frac{0.97-0.8}{\ln\frac{0.97}{0.8}} = 0.8823 \text{ atm}$$

$$k'_G \text{ for counter-diffusion} = \frac{k_G p_{BM}}{P}$$
$$= \frac{(7.11\times10^{-3})(0.8823)}{1.0}$$
$$= 6.273\times10^{-3}\ \frac{\text{kg}\cdot\text{mol}}{\text{s}\cdot\text{m}^2\cdot\text{atm}}$$
$$= 6.273\times10^{-3}\ \frac{(\text{kg})\left(\frac{2.2046\ \text{lb}}{\text{kg}}\right)(\text{mol})}{\left(\frac{\text{s}}{3600\text{s/h}}\right)(\text{m}^2)\left(\frac{10.76\ \text{ft}^2}{\text{m}^2}\right)(\text{atm})}$$
$$= 4.625\ (\text{lb mol})/(\text{h}\cdot\text{ft}^2\cdot\text{atm})$$

2.25 a. Activity coefficients are required to calculate the relative volatility.

$$\ln\gamma_1 = \frac{B/T}{\left[A\left(\frac{x_1}{x_2}\right)+1\right]^2} = \frac{-316.8/336}{\left[1.7224\left(\frac{0.575}{0.425}\right)+1\right]^2} = -0.085$$

Therefore, $\gamma_1 = 0.9185$.

$$\frac{\ln\gamma_1}{\ln\gamma_2} = \frac{1}{A(x_1/x_2)^2} \quad \text{or} \quad \ln\gamma_2 = A\left(\frac{x_1}{x_2}\right)^2 \ln\gamma_1$$

$$\ln\gamma_2 = A\left(\frac{x_1}{x_2}\right)^2 \ln\gamma_1 = 1.7224[(0.575)/(0.425)]^2(-0.085) = -0.268$$

Therefore, $\gamma_2 = 0.765$, and the relative volatility, α_{12}, can be found:

$$\alpha_{12} = \frac{\gamma_1 P_1}{\gamma_2 P_2} = \frac{0.9185\times935}{0.765\times805} = 1.39455 \approx 1.4$$

2.26 a. Use the given equilibrium diagram to determine the minimum reflux ratio. The feed to the column is 30 mol % vapor and 70 mol % liquid. Therefore, the q value $= f_l = 0.7$.

The equation for the q line is

$$y = \left(\frac{q}{q-1}\right)x - \left(\frac{x_D}{q-1}\right)$$

The slope of the q line is

$$\frac{q}{q-1} = \frac{0.7}{0.7-1} = \frac{0.7}{-0.3} = -2.334$$

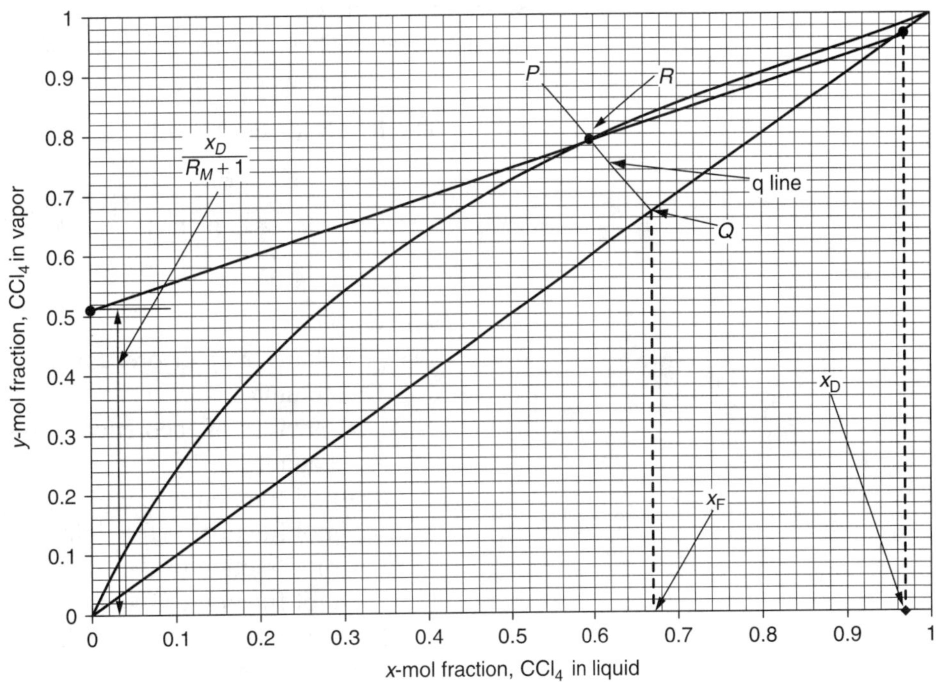

Exhibit 2.26b Diagram for estimating the minimum reflux, Problem 2.26

Draw the q line (line PQ) with a slope of -2.334 from the intersection of the vertical line at $x_F = 0.67$ and the 45° line. Line PQ intersects the equilibrium curve at point R. Draw a line from (x_D, y_D) that passes through the point R and meets the y-axis at $(x = 0,\ y = 0.51)$.

The intercept on the y-axis is 0.51, and this is equal to $x_D/(R_M + 1)$.

Therefore, $x_D/(R_M + 1) = 0.51$, from which $R_M = 0.923$ (the minimum reflux ratio).

The actual reflux ratio that will be used is $R = 2(0.923) = 1.846 \approx 1.85$.

2.27 c.

$$\begin{aligned}\text{Moles of } NH_3 \text{ in the feed} &= 100 \text{ mol}\\ \text{Moles of } NH_3 \text{ not recovered} &= 100 - 100(0.99) = 1 \text{ mol}\\ \text{Mole fraction of } NH_3 \text{ in the exit gas} &= 1/(900 + 1)\\ &= 0.00111\end{aligned}$$

The maximum liquid concentration of NH_3 in equilibrium with the feed composition is equal to x^*.

$$x^* = 0.1/1.406 = 0.0711 \text{ mol fraction}$$

$$\left(\frac{L}{G}\right)_{\min} = \frac{0.1 - 0.00111}{0.0711 - 0} = 1.391 \approx 1.4$$

The actual ratio that will be used = 1.5(1.4) = 2.1.

$$\begin{aligned}\text{Liquid rate} &= 2.1(1200) = 2520 \text{ lb mol/h}\\ &= 2520(18) = 45{,}360 \text{ lb/h}\\ \text{Cross section of tower} &= 0.7854(6)^2 = 28.27 \text{ ft}^2\end{aligned}$$

$$\text{Rate of water} = \frac{45{,}360}{500(28.27)} = 3.21 \text{ gpm/ft}^2$$

2.28 c. From the table, a sharp separation is specified between pentane and hexane. These are the key components. Pentane is the light key and hexane is the heavy key.

The Kirkbride equation is

$$\log \frac{N_R}{N_S} = 0.206 \log \left[\left(\frac{x_{HF}}{x_{LF}} \right) \left(\frac{B}{D} \right) \left(\frac{x_{LB}}{x_{HD}} \right)^2 \right]$$

where

N_R = Number of theoretical trays in the rectifying section of the column
N_S = Number of theoretical trays in the stripping section of the column
x_{HF} = mol fraction of the heavy key component in the feed
x_{LF} = mol fraction of the light key component in the feed
x_{LB} = mol fraction of the light key in the bottoms
x_{HD} = mol fraction of the heavy key in the distillate
B = Bottoms product flow, lb mol/h
D = Distillate product flow, lb mol/h

Number of theoretical stages = 17
Number of theoretical trays in the column = 17 – 1 = 16

Therefore, $N_R + N_S = 16$.

Substituting into the Kirkbride equation,

$$\log \frac{N_R}{N_S} = 0.206 \left[\left(\frac{0.20}{0.25} \right) \left(\frac{35.25}{64.75} \right) \left(\frac{0.0355}{0.015} \right)^2 \right] = 0.07463$$

Then,
$$\frac{N_R}{N_S} = 1.1875 \quad \text{or} \quad N_R = 1.1875 N_S$$

$$1.1875 N_S + N_S = 16 \qquad \therefore N_S = \mathbf{7.3}$$

$$N_R = 16 - 7.3 = 8.7$$

Therefore, theoretically, the feed plate is 8.7.

2.29 b. The equilibrium constant K at 25°C is given by

$$\Delta G_f^0 = -RT \ln K$$

$$T = 273.15 + 25 = 298.15 \text{ K}$$

Substituting the given data,

$$-3000 = -1.987(298.15)(\ln K)$$

$$\ln K = \frac{-3000}{-1.987 \times 298.15} = 5.064$$

$$K = e^{5.064} = 158.22 \text{ at } 25°\text{C}$$

If the heat of the reaction is independent of temperature, then the equilibrium constant K_T at a given temperature can be obtained by integrating the van't Hoff relation.

$$\ln\frac{K_T}{K_1}=\frac{-\Delta H^0_{rT}}{R}\left(\frac{1}{T_2}-\frac{1}{T_1}\right)$$

$T_2 = 273.15 + 50 = 323.15$ K $\qquad K_1 = 158.22$ at 25°C

Substituting the given data in the equation:

$$\ln\frac{K_T}{158.22}=\frac{-(-23{,}990)}{1.987}\left(\frac{1}{323.15}-\frac{1}{298.15}\right)$$

$$\frac{K_{50}}{158.22}=e^{-3.132805}=0.0436$$

$$K_{50}=0.0436(158.22)=6.898392\approx 6.9$$

2.30 a. For a first-order reversible reaction with an initial product concentration equal to zero, it can be shown that

$$K_E=\frac{X_{AE}}{1-X_{AE}} \quad \text{or} \quad X_{AE}=\frac{K_E}{K_E+1}$$

At t = 50°C, $K = K_E = 6.9$.

$$X_{AE}=\frac{6.9}{6.9+1}=\frac{6.9}{7.9}=0.8734$$

Therefore, $X_{AE} = 87.34\%$.

2.31 d. The following equation relates the reaction rate constant k to the activation energy E and the frequency factor α:

$$k=\alpha e^{-E/RT}$$

Taking the base-10 logarithm of both sides of the equation,

$$\log k=\log\alpha-\frac{E}{2.303R}\left(\frac{1}{T}\right)$$

Writing similar equations for log k_1 at T_1 and log k_2 at T_2, and subtracting log k_1 from log k_2, the following equation is obtained:

$$\log\frac{k_2}{k_1}=-\frac{E}{2.303R}\left(\frac{1}{T_2}-\frac{1}{T_1}\right)$$

From which, the activation energy can be found:

$$E=-\frac{2.303(R)\log(k_2/k_1)}{1/T_2-1/T_1}$$

Using the decade method of calculation,

$$k_2 = 0.005 \quad \text{then} \quad \log 0.005 = -2.301$$

Reading from the plot, $1/T_2 = 3.027 \times 10^{-3} = 0.003027$ and $T_2 \approx 330.4$ K.

$$k_1 = 0.0005 \quad \text{then} \quad \log 0.0005 = -3.301$$

Reading from the given plot, $1/T_1 = 3.19 \times 10^{-3} = 0.00319$, so $T_1 = 313.5$ K.

$$\log\frac{k_2}{k_1} = \log\frac{0.005}{0.0005} = \log 10 = 1$$

Therefore,
$$E = -\frac{2.303\,(8.314)(1)\ \text{J/mol}\cdot\text{K}}{(0.003027 - 0.00319)/\text{K}} = 117.5\frac{\text{kJ}}{\text{mol}}.$$

E can also be determined from the slope of the equation:

$$\log k = \log\alpha - \frac{E}{2.303R}\left(\frac{1}{T}\right)$$

In this case, determine the slope of the straight line in the plot and equate it with $-E/(2.303R)$.

$$\text{Slope of the line} = \frac{-3.2-(-2.3)}{0.003175-0.003027} = \frac{-0.9}{0.000148} = -6081.1$$

Therefore,

$$E = -2.303(8.314)(-6081.1) = 116{,}436\ \text{J/mol} = 116.4\ \text{kJ/mol}$$

2.32 b. The rate equation in terms of the conversion is

$$\frac{dX_A}{dt} = k_1(1-X_A) - k_2X_A$$

At equilibrium,

$$\frac{dX_A}{dt} = 0, \quad \text{and} \quad K = \frac{k_1}{k_2} = \frac{X_{AE}}{1-X_{AE}}$$

where

K is the equilibrium constant.

The solution of the equation, in terms of X_{AE}, is

$$-\ln\left(1-\frac{X_A}{X_{AE}}\right) = \frac{1}{X_{AE}}k_1t \quad t = 8\ \text{min}$$

Therefore,

$$k_1 = 0.0577\ \text{min}^{-1}$$
$$K = \frac{k_1}{k_2} = \frac{0.667}{1-0.667} = 2$$
$$k_2 = \frac{k_1}{K} = \frac{0.0577}{2} = 0.029\ \text{min}^{-1}$$

2.33 **c.** The reaction is in the liquid phase, and therefore there is no volume change. The conversion for the first-order reaction for n mixed-flow reactors in series is given by

$$X = 1 - \frac{1}{(1 + k\tau)^n}$$

In this problem, $k = 0.158\ \text{min}^{-1}$ and $n = 4$.

$$\text{Space-time, } \tau = \frac{V}{v_o} = \frac{230}{30} = 7.67\ \text{min}$$

$$k\tau = 0.158 \times 7.67$$
$$= 1.2119 \text{ (dimensionless Damköhler number)}$$

Conversion for the first-order reaction is given by

$$X = 1 - \frac{1}{(1 + k\tau)^n} = 1 - \frac{1}{(1 + 0.158 \times 7.67)^4} = 0.9582, \text{ or } 95.82\%$$

2.34 **a.** The present approximate cost of a new unit with a 125-hp motor

$$= 20{,}000(1.25)^{0.52}$$
$$= \$22{,}460$$

The cost of the unit 2 years hence = 22,460$(1.02)^2$ = \$23,400.

The installation Lang factor for centrifugal pumps is 4. Therefore, the installed cost of the unit 2 years hence = 4 × (23,400) = \$93,600.

2.35 **c.** Glass and tantalum are readily attacked by hydrofluoric acid. Therefore, the presence of fluorides rules out the selection of these two materials. SS 317L offers good corrosion resistance against wet phosphoric acid, but it is not resistant to chloride.

Comparing the given corrosion resistances of C-276 and Alloy 59, it is apparent that Alloy 59 has a better resistance to corrosion and is likely to last about 13 times longer than C-276 under similar conditions.

Alloy 59 should, therefore, be recommended.

2.36 **b.** Rewrite the given equation in the following form:

$$\frac{V}{F + kV}\frac{dC_A}{dt} + C_A = \frac{F}{F + kV}C_{Ao}$$

In the above equation, the coefficient of the derivative term is the time constant of the reactor, and the process gain, K, is the coefficient of C_{Ao}. Therefore, the equation can be rewritten as

$$\tau\frac{dC_A}{dt} + C_A = KC_{Ao}$$

The time constant for the reactor is

$$\tau = \frac{V}{F+kV} = \frac{V/F}{1+kV/F} = \frac{1.6}{1+2(1.6)} = 0.381\text{ h}$$

2.37 c. Calculation of flow parameter:

$$\frac{L}{G}\left(\frac{\rho_v}{\rho_L}\right)^{0.5} = \frac{45{,}620}{40{,}565}\left(\frac{0.213}{52.8}\right)^{0.5} = 0.071$$

Allowable clear spacing = 24 – 4 = 20 in.

For this spacing, from a chart of the capacity factor versus the flow parameter (Perry or another source),

$$\text{Capacity factor} \approx 0.29$$

$$u = 0.29(0.9)(0.95)(0.8)\left(\frac{18}{20}\right)^{0.2}\left(\frac{52.8-0.213}{0.213}\right)^{0.5} = 3.05\text{ ft/s}$$

$$\text{Vapor flow rate} = \frac{40{,}565}{0.213\times 3600} = 52.9\text{ ft}^3/\text{s}$$

$$\text{Tower cross section} = \frac{52.9}{3.05} = 17.34\text{ ft}^2$$

$$\text{Tower diameter} = \sqrt{\frac{17.34}{0.7854}} = 4.7\text{ ft}$$

Tower diameters are chosen in 6-in. increments. Therefore, use 5 ft as the diameter.

2.38 c. For the cylindrical shell,

$$t = \frac{PR}{SE-0.6P} = \frac{115(36)}{(13{,}800\times 0.85)-(0.6\times 115)} = 0.355\text{ in.}$$

The required thickness (including the corrosion allowance) = 0.355 + 0.125 = 0.48 in. The final standard thickness = 0.5 in. Therefore, a 0.5-in. thick plate will be used.

For the elliptical head,

$$t = \frac{115\times 36}{(13{,}800\times 0.85)-(0.1\times 115)} = 0.3533\text{ in.}$$

This is smaller than the size calculated for the shell. Therefore, the thickness of the shell controls, and a thickness of 0.5 in. is adequate.

MAWP is calculated on the basis of the thickness excluding the corrosion allowance. Therefore,

$$\text{MAWP} = \frac{SEt}{R+0.6t} = \frac{13{,}800\times 0.375}{36+0.6\times 0.375} = 149\text{ psig}$$

2.39 b. According to the API 2000, 75% of the total area, or the total exposed surface area, up to a height of 30 ft, is to be taken as the wetted area. In the present case, the total area needs to be taken since the vessel is installed within a 30-ft height above the ground.

Calculation of area:

$$\text{Area of cylindrical shell} = \pi DL = \pi(5)(10) = 157.1 \text{ ft}^2$$

$$\text{Area of two 2:1 elliptical heads} = 2(1.19)(5)^2 = 59.5 \text{ ft}^2$$

$$\text{Total wetted area} = 157.1 + 59.5 = 216.6 \text{ ft}^2$$

For $200 < A < 1000$ ft^2, the applicable equation is $199{,}300(F)(A)^{0.566}$.

For a bare vessel, $F = 1$.

Therefore, $Q_F = 199{,}300(216.6)^{0.566} = 4{,}183{,}020$ Btu/h $= 4.18$ MMBtu/h.

2.40 d.

$$\%\text{ BOD oxidized} = \frac{250}{350} \times 100 = 71.43\%$$

$$\%\text{ BOD remaining} = 100 - 71.43 = 28.57\%$$

$$L_T = L(10^{-kt})$$

$$\text{From which, } \log\frac{L_T}{L} = -kt$$

When $t = 5$ days, % BOD remaining = 28.57%. Therefore,

$$\log\frac{0.2857 \times 350}{350} = -k(5)$$

$$k = (-0.5441)/(-5) = 0.109 \text{ at } 20°\text{C}$$

The time required to satisfy the 50%-O_2 demand is given by

$$\log 0.5 = -0.109t$$

$$t = 2.76 \text{ days} \approx 2.8 \text{ days}$$

SOLUTION AND TOPIC SUMMARY—MORNING EXAM

Problem	Solution	Exam Topic Category	Subtopic
		Mass/Energy Balances and Thermodynamics	
1.1	c		Fuels and combustion
1.2	d		Fuels and combustion
1.3	a		Mass balance
1.4	b		Mass balance
1.5	a		Mass balance
1.6	c		Energy balance
		Mass/Energy Balances and Thermodynamics	
1.7	a		Energy balance
1.8	b		Property evaluation
1.9	b		Work and energy
1.10	c		Heat of vaporization
		Fluids	
1.11	c		Bernoulli equation
1.12	c		Control valves
1.13	c		Pipe schedule and roughness
1.14	d		Reynolds number
1.15	b		Centrifugal pump
1.16	b		Fluid measurement, pitot tube
1.17	c		Fluid friction, pump head
		Heat Transfer	
1.18	a		Resistances in series
1.19	b		Convective heat transfer
1.20	c		Overall heat-transfer coefficient
1.21	a		Mean temperature difference
1.22	d		Overall heat-transfer coefficient
1.23	b		Heat loss from pipe
		Mass Transfer	
1.24	a		Molecular diffusion
1.25	b		Bubble-point temperature
1.26	a		Number of theoretical stages
1.27	b		Packed column height, N_{OG}
1.28	a		Packed column diameter
		Chemical Reaction Engineering	
1.29	b		Rate constant
1.30	a		Activation energy
1.31	b		Percent conversion, CSTR
1.32	a		Plug-flow reactor conversion
1.33	a		Product distribution, parallel reactions

(Continued)

Problem	Solution	Exam Topic Category	Subtopic
		Plant Design and Operation	
1.34	b		Time value of money
1.35	c		Corrosion
1.36	b		Process control
1.37	b		Plant safety
1.38	b		Plant safety
1.39	c		Safety relief valve
1.40	a		Equipment design

SOLUTION AND TOPIC SUMMARY—AFTERNOON EXAM

Problem	Solution	Exam Topic Category	Subtopic
		Mass/Energy Balances and Thermodynamics	
2.1	b		Fuels and combustion
2.2	c		Stoichiometry
2.3	b		Mass balance, recycling
2.4	a		Mass balance, distillation column
2.5	b		Phase behavior
2.6	b		Mass/energy balance
2.7	c		Energy balance, heating value of fuel
2.8	d		Mass balance, use of steam tables, heat of vaporization
2.9	b		Heat of vaporization
2.10	a		Fuels, heat of combustion
		Fluids	
2.11	c		Fanning friction factor
2.12	a		Velocity and pressure drop in a pipeline
2.13	d		Centrifugal pump, suction pressure
2.14	b		Pump bhp and mhp
2.15	b		Flow in pipes
2.16	c		Control valve
2.17	b		Fluid friction, pressure drop
		Heat Transfer	
2.18	d		Prandtl number, convection
2.19	b		Conduction heat transfer
2.20	c		Heat loss from a pipe
2.21	b		Natural convection
2.22	a		Condensation
2.23	b		Convection heat transfer
		Mass Transfer	
2.24	d		Mass transfer coefficients
2.25	a		Vapor-liquid equilibrium
2.26	a		Determination of minimum reflux, McCabe-Thiele method
2.27	c		Packed column, minimum rate of solvent
2.28	c		Feed plate location
		Chemical Reaction Engineering	
2.29	b		Equilibrium constant
2.30	a		Equilibrium conversion
2.31	d		Activation energy
2.32	b		Reaction rate constant
2.33	c		CSTRs in series

(*Continued*)

Problem	Solution	Exam Topic Category	Subtopic
		Plant Design and Operation	
2.34	a		Equipment cost index
2.35	c		Corrosion, material of construction
2.36	b		Process control
2.37	c		Equipment design
2.38	c		Equipment design
2.39	b		Plant safety, fire heat load
2.40	d		Environmental

CHAPTER 5

Solutions to Additional Review Problems

OUTLINE

MASS/ENERGY BALANCES AND THERMODYNAMICS

3.1 **d.** The overall reaction is

$$\underset{2(131.5)}{2CHCLCl_2} + \underset{32}{O_2} \rightarrow \underset{2(147.5)}{2CHCl_2COCl}$$

$$\text{TCE required} = \frac{2\times 131.5}{2\times 147.5}\times 100 = 89.2 \text{ lb/100 lb DCAC}$$

$$\text{TCE actually used} = \frac{89.2}{0.849} = 105 \text{ lb/per 100 lb of DCAC}$$

3.2 **b.** O_2 theoretically required $= \dfrac{32}{2 \times 147.5} \times 1000 = 108.5$ lb

O_2 actually required = (108.5/0.603) = 180 lb/1000 lb DCAC

3.3 **c.** Basis = 100 mol of CH_4

Assume complete combustion.

$$CH_4 + 2O_2 = CO_2 + 2H_2O$$

	Feed mol		Products mol	O_2 Used mol
CH_4	90	CO_2	90	90
N_2	10	H_2O	180	90

Let x be mol of oxygen used. Then nitrogen in air = (0.79/0.21)(x) mol

Total nitrogen in flue gas = 10 + 3.7619x mol

Unreacted oxygen remaining = (x – 180) mol

Then, taking ratios,

$$\frac{x-180}{10+3.7619x} = \frac{4.54}{86.38}$$

Simplification yields 86.38x – 15548.4 = 45.4 + 17.079x

Solving for x:

$x = 225$ mol of oxygen used

Excess air:

Excess O_2 = 225 – 180 = 45 mol

$$\text{Excess air} = \frac{45}{180} \times 100 = 25\%$$

3.4 **a.** Overall balance: $100 = M + P$

where
M = moles of NH_3 produced
P = purge, mol per hour

Overall H_2 balance: $100(0.751) = M(0) + P(0.8)$

From the second equation, $P = \dfrac{100(0.751)}{0.8} = 93.875$ mol/h

From the first equation $M = 100 - 93.875 = 6.125$ mol/h of NH_3

Formation of 1 mol of NH_3 requires 1.5 mol of H_2

Mol of hydrogen consumed = 6.125(1.5) = 9.1875 mol/h

Percent conversion of hydrogen per pass $= \dfrac{9.1875}{75.1} \times 100 = 12.23\%$

3.5 **d.** First, calculate the molecular wt and mol of feed.

Basis of calculation: 1 kg mol

Component	Mol Fraction	Mol wt.	Contribution to mol wt.
C_6H_{14}	0.20	86	17.2
N_2	0.80	28	22.4

mixture molecular wt. = 39.6

Mol of feed = (10000/39.6) = 252.53 kg mol

Hexane in feed = 252.53(0.2) = 50.51 kg mol

N_2 in feed = 252.53(0.8) = 202.02 kg mol

$$\text{Hexane in gas stream leaving the condenser} = \frac{0.05}{0.95} \times 202.02$$
$$= 10.63$$

By material balance on hexane, hexane condensed = 50.51 – 10.63
= 39.88 kg mol

Pounds of hexane recovered per hour = 39.88(86)
= 3429.7 lb say 3430 lb/h

3.6 **a.** Vapor pressure of water at 30°C = 31.82 mm Hg from steam tables

Mol fraction, y_{H_2O} = (31.82/740) = 0.043

$$\text{Volume of feed at standard conditions} = 1500\left(\frac{273}{273+30}\right)\left(\frac{740}{760}\right)$$

= 1316 ft^3 = (1316/359)
= 3.67 mol of wet air

Therefore, the amount of water condensed = (3.67 × 0.043 × 0.75) (18)
= 2.13 lb

3.7 **b.**

Conversion of lb into lb mol

Component	lb	Mol wt.	lb mol
CO_2	55	44	1.25
N_2	140	28	5.0
O_2	32	32	1.0
			7.25

T = 460 + 120 = 580°R

$$P = \frac{nRT}{V} = \frac{7.25 \times 10.73 \times 580}{1000} = 45.1 \text{ psia} = 30.4 \text{ psig}$$

3.8 **c.** Calculate the reduced parameters first.

$$T = 273 + 60 = 333 \text{ K}$$

$$P_r = \frac{24}{115.5} = 0.215 \qquad T_r = \frac{333}{405.4} = 0.82$$

From z chart at $P_r = 0.215$ and $T_r = 0.82$

$$z = 0.842$$

$$R = \frac{82.06 \text{ cm}^3 \times \frac{10^{-6} \text{ m}^3\text{atm}}{\text{cm}^3}}{\text{g mol} \cdot \text{K}}$$

$$\text{g mol of gas in tank} = n = \frac{PV}{zRT} = \frac{24(2)}{0.842 \times 82.6 \times 10^{-6} \times 333}$$
$$= 2086.2 \text{ g mol}$$

$$= \frac{2086.2 \times 17}{1000} = 35.47 \text{ kg} \approx 35.5 \text{ kg}$$

3.9 **c.** Write the energy balance for the two steps separately.

Step 1: Heating of gas. The energy balance reduces to

$$\Delta U = Q$$

since there is no mass flow in and out, and no work done, it is a closed system.

Then $$Q = 2 \text{ kcal} = \frac{\text{kcal}\, 10^3 \text{ cal}}{\text{kcal}} \left| \frac{1 \text{ J}}{0.23901 \text{ cal}} \right. = 8370 \text{ J}$$

Step 2: Expansion of gas

In this step the gas expands and does work on the piston that moves. Heat is also added to the gas to maintain the temperature. The energy balance for the closed system reduces to

$$0 = Q + W$$

$$\Delta U = 0 \quad \text{as } \Delta T = 0\text{, no mass flow in and out}$$

Then $$Q = -W = -(-100 \text{ J}) = 100 \text{ J}$$

The total heat transferred to gas in 2 steps = 8370 + 100 = 8470 J

3.10 **b.** Get enthalpies of feed water streams and the saturated steam product from steam tables.

Liquid water at 80°F, $h_l = 48.02$ Btu/lb

Liquid water at 150°F, $h_l = 117.87$ Btu/lb

Saturated steam at 230.3 psig, $H_v = 1200.9$ Btu/lb

For convenience, calculate enthalpy of mixed water feed stream. Assume c_P of water = 1.

$$\text{Then enthalpy of mixed stream} = \frac{16,500 \times 48.02 + 23,500 \times 117.87}{40,000}$$

$$= 76.83 \text{ Btu/lb}$$

The energy balance for the open system reduces to

$$Q = \Delta H + \Delta \text{KE}$$

(Flow process PE change is negligible. Inlet velocity is not significant, no work involved.)

$$\Delta H = M(1200.9 - 76.83) = 40,000(1124.07) = 44,962,800 \text{ Btu/h}$$

$$\Delta \text{KE} = \frac{1}{2}\frac{(40,000)(630^2)}{32.2 \times 778} = 3.17 \times 10^5 \text{ Btu/h}$$

Total heat to be supplied to the boiler = $44.9628 \times 10^6 + 3.17 \times 10^5$
= 45.3×10^6 Btu/h
= 45.3 MMBtu/h

3.11 b. The energy balance on the closed system reduces to

$$Q = \Delta H$$

$$H = U + PV \qquad \therefore\ U = H - PV$$

and

$$\Delta U = \Delta H - \Delta PV$$

$$\Delta H = 418.2 - 180.29 = 237.91 \text{ Btu/lb}$$

$$\Delta PV = \left[1.262\frac{\text{ft}^3}{\text{lb}} \times (80.11)\text{psia} - 0.0232\frac{\text{ft}^3}{\text{lb}} \times 2.867\text{psia}\right]$$

$$\times \frac{1.987}{10.73}\frac{\text{Btu}}{\text{ft}^3 \cdot \text{psia}} = 18.71 \text{ Btu/lb}$$

Therefore, $\Delta U = \Delta H - \Delta PV = 237.91 - 18.71 = 219.2$ Btu/lb

3.12 d. An energy balance on the condenser yields $Q = \Delta H$

Average molecular wt of product = 0.97(78) + 0.03(92)
= 78.42 kg/kg mol.

Mol of distillate = 5500/78.42 = 70.14 mol/h

Vapor flow rate from the column to the condenser = V = 2 × 70.14
+ 70.14
= 210.42 mol/h

Average latent heat of distillate = 0.97 × 30.76 + 0.03 × 33.5
= 30.84 kJ/g mol
= 3.084×10^4 kJ/kg mol

$$\text{Therefore, condenser duty} = Q = 3.084 \times 10^4 \frac{\text{kJ}}{\text{kg mol}} \times 210.42 \frac{\text{kg mol}}{\text{h}}$$

$$= 6.49 \times 10^6 \text{ kJ/h}$$

3.13 **a.** Calculate the heat of vaporization at the boiling point by Chen's equation.

$$\Delta H_{vBP} = \frac{T_B[0.033(T_B/T_C) - 0.0327 + 0.0297 \log P_C]}{1.07 - (T_B/T_C)}$$

$$T_B = 273.1 + 38.2 = 311.2 \text{ K}$$

$$\Delta H_{vBP} = \frac{311.3[0.033(311.3/504) - 0.0327 + 0.0297 \log 61.5]}{1.07 - (311.3/504)}$$

$$= 28.1 \text{ kJ/g mol}$$

$$\Delta H_{V(0°\text{C})} = \Delta H_{V(38.2°\text{C})} \left[\frac{T_C - T}{T_C - T_B} \right]^{0.38}$$

$$= 28.1 \left[\frac{504 - 273.1}{504 - 311.3} \right]^{0.38} = 30.1 \text{ kJ/g mol}$$

Therefore, the heat of vaporization per kg mol = 30.1×10^3 kJ/kg mol

3.14 **b.** From steam tables, at 70°F, $h_L = 38.04$ Btu/lb

Again from steam tables, at 260 psia and 1500°F,

$$H_V = 1800.1 \text{ Btu/lb}$$

Enthalpy change = 10(1800.1 – 38.04) = 17,620.6 Btu

3.15 **a.** Mass balance reduces to $\sum M_I = \sum M_o$

Energy balance reduces to $W = M_S \left(\sum \hat{H}_o - \sum \hat{H}_I \right)$

where
H = enthalpy, Btu/lb and
M_s = rate of steam, lb/h

Since the process is adiabatic, $Q = 0$, and $\hat{S}_I = \hat{S}_o$ where S = entropy, Btu/lb · °F

At 150 psia and 900°F, $\hat{S}_I = \dfrac{1.8301 + 1.8451}{2} = 1.8376$ Btu/lb · °F

$\hat{H}_I = 1477.85$ Btu/lb

At 20 psia, $\hat{S}_o = 1.8376$ Btu/lb · °F

Calculate t:

$$t = \frac{1.8376 - 1.7808}{1.8396 - 1.7808} \times (400 - 300) + 300 = 396.6\,°\text{F}$$

$$\hat{H}_o = \frac{396.6 - 300}{400 - 300} \times (1239.2 - 1191.6) + 1191.6 = 1237.97 \text{ Btu/lb}$$

Actual work $W = 0.85(1237.97 - 1477.85) = -203.9$ Btu/lb

The minus sign now indicates that the work is done by the system.

$$\text{Therefore turbine output} = \frac{M_s|W|}{2545} = \frac{20,000 \times 203.9}{2545} = 1602.4 \text{ hp}$$

3.16 **c.** $P = 100$ psia. The throttling process is isenthalpic. Therefore,

$$\hat{H}_o = 1477.85 \text{ Btu/lb}$$

$$t = \frac{1477.85 - 1428.9}{1479.5 - 1428.9} \times (900 - 800) + 800 = 896.7°\text{F}$$

$$\hat{S}_I = 0.9674(1.8829 - 1.8443) + 1.8443 = 1.8816 \text{ Btu/lb} \cdot °\text{F}$$

For maximum work, $\hat{S}_o = 1.8816$ Btu/lb·°F

$$t_o = \frac{1.8816 - 1.8396}{1.8918 - 1.8396} \times 100 + 400 = 480.5°\text{F}$$

$$\hat{H}_o = \frac{480.5 - 400}{500 - 400}(1286.6 - 1239.2) + 1239.2 = 1277.36 \text{ Btu/lb}$$

At 80% efficiency actual work = 0.8(−1477.85 + 1277.36)
= −160.4 Btu/lb

$$\hat{H}_o = 1477.85 - 160.4 = 1317.45 \text{ Btu/lb}$$

$$t_o = \frac{1317.45 - 1286.6}{1334.4 - 1286.6}(600 - 500) + 500 = 565°\text{F}$$

3.17 **d.** For minimum work, the process is reversible and isentropic and

$$\hat{S}_I = \hat{S}_o = 1.4896 \text{ Btu/lb} \cdot °\text{R}$$

Calculate outlet enthalpy and temperature. From the properties at 60 psia,

$$t_o = 240 + \frac{1.4896 - 1.4819}{1.4976 - 1.4819}(20) = 249.8 \cong 250°\text{F}$$

$$\hat{H}_o = 739.7 + \frac{1.4896 - 1.4819}{1.4976 - 1.4819}(750.9 - 739.7) = 745.2 \text{ Btu/lb of } NH_3$$

Therefore, isentropic reversible work is by energy balance,

$$W = \hat{H}_o - \hat{H}_I = 745.2 - 643.6 = 101.6 \text{ Btu/lb of ammonia}$$

The positive sign indicates that the work is done on the system. If the compressor efficiency is 80%, the actual work will be as follows:

Actual work = 101.6/0.8 = 127 Btu/lb of NH_3

3.18 **b.** Actual work $= \hat{H}_o - \hat{H}_I = 127$ Btu/lb of ammonia calculated in Problem 3.17

Therefore, $\hat{H}_o = \hat{H}_I + 127 = 770.6$ Btu/lb of NH_3

From given table interpolate temperature as follows:

$$t_o = 280 + \frac{770.6 - 762.1}{773.3 - 762.1} \times 20 = 295.2°F$$

$$\text{Entropy at outlet} = \hat{S}_o = 1.513 + \frac{15.2}{20}(1.5281 - 1.513)$$
$$= 1.5245 \text{ Btu/lb}\cdot°F$$

Entropy increase during the irreversible process = 1.5245 – 1.4896
= 0.0349 Btu/lb · °F

3.19 **c.** First calculate heat of reaction at constant pressure as follows:

$$\Delta H_r = \sum(\Delta H_f)_{\text{products}} - \sum(\Delta H_f)_{\text{reactants}} = -166.2 + 235.31$$
$$= 69.11 \text{ kJ/mol}$$

Now $$\Delta U_r = \Delta H_r - RT\left(\sum_{\substack{\text{gaseous}\\\text{products}}} v_i - \sum_{\substack{\text{gaseous}\\\text{reactants}}} v_i\right)$$

Products: $\sum v_i = 1 + 1 = 2$

Reactants: $\sum v_i = 1$

Then $$\Delta U_r = 69.11 - \frac{8.314\text{ J}}{\text{mol}\cdot\text{K}}\left|\frac{298\text{ K}}{1}\right|\frac{1\text{ kJ}}{10^3\text{ J}}$$

$$= 69.11 - 2.48 = 66.63 \frac{\text{kJ}}{\text{mol}}$$

3.20 **c.** The heating value is the negative value of the heat of combustion. First calculate the lower heating value as in the following table

Component	Mol %	Mol wt.	Wt in lb per mol of Gas	Heating Value Btu/lb mol	Heating value of Component in Gas Btu/lb mol Gas	Mol of Water Formed in Combustion per mol of Gas
CH_4	0.80	16	12.8	345,176	276,141	1.6
C_2H_6	0.15	30	4.5	615,672	92,351	0.45
C_3H_8	0.05	44	2.2	887,464	44,373	0.2
			19.5		412,865	2.25

HHV per lb mol of fuel gas = 412,865 + (2.25 × 18.02 × 1053)
= 455,559 Btu/lb mol of fuel gas
HHV per pound of fuel + (455,559/19.5) = 23,362 Btu/lb of fuel gas

3.21 **b.** Calculate heat of vaporization at 25°C:

heat of vaporization at 25°C = 273.1 + 25 = 298 K

Using Watson's relation

$$T_{r2} = \frac{298}{699} = 0.4263 \qquad T_{r1} = \frac{486.6}{699} = 0.6961$$

$$\Delta H_{V298} = \left(\frac{1-0.4263}{1-0.6961}\right)^{0.38} \times 11,035 = 14,049 \text{ cal/g mol}$$

Entropy change from saturated liquid to saturated vapor:

$$\Delta S^{sat} = \frac{14,049}{298} = 47.1 \text{ cal/(g mol}\cdot\text{K)} = 47.1 \text{ (Btu/lb mol}\cdot{}^\circ\text{R)}$$
$$= 47.1/150.177 = 0.314 \text{ Btu/(lb}\cdot{}^\circ\text{R)}$$

3.22 **d.** Watson's equation:

$$\Delta H_{v2} = \Delta H_{v1}\left(\frac{1-T_{r2}}{1-T_{r1}}\right)^{0.38}$$

$$T_{r1} = \frac{273.15+97.2}{273.15+263.7} = \frac{370.35}{536.85} = 0.690$$

$$T_{r2} = \frac{273.15+115}{273.15+263.7} = \frac{388.15}{536.85} = 0.723$$

$$\Delta H_{v2} = 164.36\left[\frac{1-0.723}{1-0.69}\right]^{0.38} = 157.5 \text{ cal/g}$$

$$= 157.5(5.1868)(60.09)$$
$$= 39,625 \text{ J/g mol}$$

3.23 **b.** Residual volume is the difference between the volume of ideal gas and the actual volume at the same conditions of temperature and pressure.

Actual volume per lb mol = 0.1296(64) = 8.2944 ft^3/lb mol

Volume of SO_2 gas at the same conditions if it were an ideal gas at these conditions

$$\frac{RT}{P} = \frac{10.73(460+480)}{1000} = 10.0862 \text{ ft}^3\text{/lb mol}$$

$$\text{Residual volume} = \frac{RT}{P} - \hat{V} = 10.0862 - 8.2944 = 1.7918 \text{ ft}^3\text{/lb mol}$$

$$= \frac{1.7918}{64} = 0.028 \text{ ft}^3\text{/lb}$$

3.24 **d.** From the tables, at 150 psia,

$$t = 105.1°\text{F}, \quad \hat{h} = 46.9 \text{ Btu/lb}, \quad \hat{S}_l = 0.095 \text{ Btu/lb·°F}$$

For maximum work, entropy change is zero. Therefore entropy at outlet of turbine = 0.095 Btu/lb·°F for maximum work.

At 0°F, $\hat{H}_v = 103.1$ Btu/lb, $\hat{h}_l = 12.1$ Btu/lb,

$\hat{S}_v = 0.2255$ Btu/lb·°F, $\hat{S}_l = 0.0275$ Btu/lb·°F

At turbine exit if x is the quality of vapor, then

$$0.2255x + 0.0275(1-x) = 0.095$$

Therefore, $x = 0.0675/0.1980 = 0.341$ (vapor) and 0.659 mass fraction liquid.

Enthalpy at turbine exit = 0.341(103.1) + 12.1(0.659) ≈ 43.1 Btu/lb

$$\text{Maximum work} = 43.1 - 46.9 = -3.8 \text{ Btu/lb}$$

$$\text{Actual work} = (0.85)(-3.8) = -3.23 \text{ Btu/lb}$$

Actual enthalpy at turbine exit = 46.9 – 3.23 = 43.67 Btu/lb

Actual quality of vapor at turbine exit: $x(103.1) + (1-x)(12.1) = 43.67$

From which $x = 0.347$ and $(1-x) = 0.653$ mass fraction liquid.

Actual exit entropy is given by

$$0.347(0.2255) + 0.653(0.0275) = 0.0962 \text{ Btu/lb·°F}$$

Therefore, entropy increase = 0.0962 – 0.095 = 0.0012 Btu/lb·°F

3.25 **d.**

$$\Delta S = \int_{460+68}^{460+94.3} \frac{0.521dT}{T} + \frac{151.06}{460+94.3} + \int_{460+94.3}^{460+122} \frac{0.44dT}{T} \quad \text{at constant pressure}$$

$$= 0.521 \ln\frac{554.3}{528} + \frac{151.06}{554.3} + 0.44 \ln\frac{582}{554.3}$$

$$= 0.3193 \text{ Btu/lb·°R}$$

Therefore, entropy change per lb mol = 0.3193(74.12) = 23.67 Btu/(lb mol·°R)

3.26 **c.** Energy balance for the conditions of the problem with the vessel as the system reduces to

$$U_E - U_B = Q$$

Steam tables do not give U values for superheated steam. By definition,

$$H = U + PV$$

Therefore, $\Delta U = \Delta H - \Delta(PV)$

$$\text{Initial volume} = \text{final volume} = 1.3044 \text{ ft}^3$$

Let x be the quality of steam initially. Then

$$V_1 = 3.015x + 0.01809(1 - x) = 1.3044 \quad \text{hence, } x = 0.4292$$

$$\hat{H}_B = 1194.1(0.4292) + 330.51(1 - 0.4292) = 701.2 \text{ Btu/lb}$$

$$Q \text{ (heat added)} = (1357 - 701.2) - \frac{(500-150)(1.3044)(144)}{778}$$
$$= 1357 - 701.2 - 84.5 = 571.3 \text{ Btu/lb}$$

3.27 **a.** From the second law of thermodynamics,

$$\frac{W}{Q_H} = \frac{T_H - T_L}{T_H} = \frac{535-490}{535} = 0.08411$$
$$W = 0.08411(200{,}000) = 16{,}822 \text{ Btu/h}$$

$$Q_L = \frac{W}{\frac{T_H}{T_L} - 1} = \frac{16{,}822}{\frac{535}{490} - 1} = 183{,}173 \text{ Btu/h}$$

3.28 **c.** The critical constants of ethylene are $T_c = 282.7$ K and $P_c = 50.9$ atm.

$$PV = ZnRT$$

$$T_c = 282.7 \text{ K} = 282.7(1.8) = 508.9°\text{R}$$

$$P = 500 \text{ atm} \qquad P_r = \frac{500}{50.9} = 9.823 \qquad T_r = \frac{383.1}{282.7} = 1.36$$

From the high pressure Z chart, $Z \cong 1.15$

$$n = \frac{PV}{ZRT} = \frac{500(2)}{1.15(0.732)(508.9)} = 2.3343 \text{ lb mol}$$
$$= 2.3343(28) = 65.4 \text{ lb}$$

3.29 **c.** $2 \text{ hp} = 2(2545) = 5090$ Btu

By heat balance, $W = Q_h - Q_L$

$$\therefore Q_h = 5090 + 7500 = 12{,}590 \text{ Btu/h}$$

From the second law of thermodynamics,

$$\frac{Q_h}{Q_L} = \frac{T_h}{T_L} = \frac{T_h}{520} = \frac{12{,}590}{7500} = 1.6787$$

$$\therefore T_h = 1.6787(520) = 873°R = 413°\text{F}$$

3.30 **c.** Evaluate $\int_1^2 PdV$ for one mole of an ideal gas, $PV = RT$

$$\therefore P = \frac{RT}{V}$$

Therefore $\displaystyle\int_1^2 PdV = \int_1^2 \frac{RT}{V} dV = RT \ln \frac{V_2}{V_1} = W$

FLUID MECHANICS

3.31 **b.** Applying the Bernoulli equation to points A and B,

$$\frac{P_A}{\rho_A} + Z_A + \frac{u_A^2}{2g_c} - F + w = \frac{P_B}{\rho_B} + Z_B + \frac{u_B^2}{2g_c}$$

$Z_B = 0$ taken as datum plane, $w = 0$ no pump in line

The Bernoulli equation reduces to

$$\frac{P_A}{\rho_A} + Z_A + \frac{u_A^2}{2g_c} - F = \frac{P_B}{\rho_B} + \frac{u_B^2}{2g_c}$$

or

$$\frac{P_A - P_B}{\rho_A} = F - Z_A - \left(\frac{u_A^2 - u_B^2}{2g_c}\right)$$

$$F = \frac{1.34}{12} \times \frac{1.05 \times 62.4}{0.0754} = 97.03 \text{ ft of air}$$

$$Z_A = 10 \text{ ft (given)} \quad \frac{u_A^2 - u_B^2}{2g_c} = \frac{2.9^2 - 0.28^2}{64.4} = 0.13 \text{ ft}$$

$$\frac{P_A}{\rho_A} = \frac{P_o \times 144 + \frac{6}{12} \times 62.4 \times 2.5}{0.0754} = \frac{P_o \times 144}{0.0754} + 1034.5 \text{ ft of air}$$

$$\frac{P_B}{\rho_A} = \frac{P_o \times 144 + h_l \times 62.4 \times 1.05}{0.0754} = \frac{P_o \times 144}{0.0754}$$

$$+ \frac{h_l \times 62.4 \times 1.05}{0.0754} \text{ ft of air}$$

$$\frac{P_A - P_B}{\rho_A} = 1034.5 - \frac{h_l \times 62.4 \times 1.05}{0.0754} = 97.03 - 10 - 0.13$$

From which,

$$h_1 = 1.0905 \text{ ft} = 13.09''$$

Therefore, the height from the bottom of the tank = 13.09 + 3 ≈ 16.1".

3.32 **c.**

$$\Delta P = \frac{2fLu^2\rho}{g_c d_i (144)}$$

Since both f and u are unknown, the solution is by trial and error. For the first trial assume $u = 4$ ft/s.

$$\frac{Du\rho}{\mu} = \frac{0.1723 \times 4 \times 51}{2.2 \times 0.000672} = 2.38 \times 10^4 \quad \text{Relative roughness} = 0.0009$$

$f = 0.00675$ from the Fanning friction chart

$$\Delta P = \frac{2 \times 0.00675 \times 400 \times 4^2 \times 51}{32.2 \times 0.17225 \times 144}$$

$$= 5.52 \text{ psi}$$

This is <7.5 psi required.

Assume $u = 4.66$ ft/s [$7.5/5.52 = (u_2/u_1)^2$ from which, $u_2 = 4.66$ ft/s]

$$\frac{Du\rho}{\mu} = \frac{4.66}{4.0} \times 2.38 \times 10^4 = 2.8 \times 10^4 \qquad f = 0.00675 \text{ approx.}$$

$$\Delta P = \frac{0.0675 \times 400 \times 4.66^2 \times 51}{32.2 \times (0.1723) \times 144} = 7.49 \text{ psi}$$

which is very close to the required value. Therefore $u = 4.66$ ft/s $\doteq$ 4.7 ft/s

3.33 **b.** Assume turbulent flow. Therefore

$$C_O = 0.61.$$

The pitot tube positioned at the center gives the maximum velocity.

$$\text{Density of air} = \frac{29}{359} \times \frac{492}{680} \times \frac{750}{760} = 0.05768 \text{ lb/ft}^3$$

$$4\text{" of water column} = \frac{4}{12} \times \frac{62.4}{0.05768} = 360.6 \text{ ft of air}$$

$$u_{\max} = 0.98\sqrt{64.4 \times 360.6} = 149.3 \text{ ft/s}$$

Average velocity u = 0.82(149.3) = 122.4 ft/s

$$\text{Reynolds number} = \frac{Du\rho}{\mu} = \frac{(20/12)(122.4)(0.05768)}{0.022(0.000672)} = 7.96 \times 10^5$$

Area of cross section = $0.7854(20/12)^2 = 2.182$ ft^2

Flow = 2.182(122.4)(60) = 16,025 cfm at 750 mm and 220°F

Therefore, the flow at 60°F and 760 mm pressure,

$$= 16{,}025\left(\frac{520}{680}\right)\left(\frac{750}{760}\right) = 12{,}093 \text{ scfm}$$

3.34 **c.** Losses in fittings, contraction, etc. are given in terms of resistance coefficients K. L is the straight length of pipe and does not include equivalent length of fittings.

Expressions a and d calculate head loss only due to straight pipe and do not include the loss due to fittings. This excludes answers a and d. Four times f' is used in expression b. The head loss due to straight pipe will be four times the actual head loss if the Moody friction factor f' is multiplied by 4. This excludes expression b.

3.35 **b.**

$$\text{Velocity } u \text{ in 2" pipe} = \frac{40}{7.48 \times 60 \times 0.0233} = 3.83 \text{ ft/s}$$

$$\frac{Du\rho}{\mu} = \frac{0.1723 \times 3.83 \times 62.3}{0.982 \times 0.000672} = 6.23 \times 10^4$$

$\in/D = 0.00015/0.1723 = 0.00087$

$f = 0.00575$ from the Fanning friction chart

3.36 **d.** $K = 0.78$ for 3" Schedule 40 pipe since liquid enters this pipe. To express in terms of 2" pipe, use the following relation:

$$K_a = K_b\left(\frac{d_a}{d_b}\right)^4$$

where

K_a, K_b = Resistance coefficients for 2" and 3" std. pipes
d_a, d_b = internal diameters of 2" and 3" std. pipes

$$\text{Then } K_a \text{ for 2" pipe} = K_b\left(\frac{d_a}{d_b}\right)^4 = 0.78\left(\frac{2.067}{3.068}\right)^4 = 0.161$$

3.37 **c.** Flow through $1^1/_2$" pipe is 20 gpm.

$$u = \frac{20}{7.48 \times 60 \times 0.01415} = 3.15 \text{ ft/s}$$

$$\text{Re No.} = \frac{Du\rho}{\mu} = \frac{0.1342 \times 3.15 \times 62.3}{0.982 \times 0.000672} = 3.99 \times 10^4$$

3.38 **c.** Velocity through the orifice $u_o = \dfrac{34.1}{3600 \times 0.001553} = 6.1$ m/s

$$\frac{Du_o p}{\mu} = \frac{0.0445(6.1)(1000)}{1 \times 10^{-3}} = 2.72 \times 10^5$$ (Figure 5-18, Perry's Handbook)

Therefore, flow is fully turbulent.

Hence, $C_O = 0.61$ $\beta = 1.75/3.068 = 0.5704$ $1 - \beta^4 = 0.8941$

Hence, $6.1 = 0.61\sqrt{\dfrac{2(9.81)(\Delta H)}{0.8041}}$

Then $\Delta H = 4.56 \text{ m} = 456 \text{ cm}$

$$\Delta H = h_m\left(\frac{13.6}{1} - 1\right)$$

$$h_m = \Delta H/12.6 = 456/12.6 = 36.2 \text{ cm}$$

3.39 **d.** To calculate TDH, pump suction pressure is needed. So calculate the pump suction pressure first.

Operating pressure = 0.0 psig

Static head = (26 – 3)(0.433)(0.81)

= 8.1 psi

Assume the suction line loss = –1.5 psi

Suction pressure = 6.6 psig

TDH = 190 – 6.6 = 183.4 psi
= 183.4 × 2.31/0.81 = 523.03 ft

3.40 **c.** From Problem 3.39, suction pressure = 6.6 psig

From given data, vapor pressure = 0.0 psig

$$\begin{aligned} NPSH_A &= 6.6 \text{ psi} \\ &= 6.6 \times 2.31/0.81 \\ &= 18.8 \text{ ft} \end{aligned}$$

Usually the calculated $NPSH_A$ is reduced by 2 ft for reporting on the specification sheet to account for uncertainty in the calculated value. Thus, in the present case the reported value will be 16.8 ft.

3.41 **b.** Calculation of maximum suction pressure: Maximum suction pressure is based on the design pressure of vessel and maximum operating level in the tank or HLL.

Origin pressure = 50.0 psig.

Static head = [(26 − 3) + 6 +10/12)] × 0.433 × 0.81 = 10.5 psi

−line loss = −1.5 psi

Maximum suction pressure = 59.00 psig

3.42 **c.** Resistance coefficient $= K = \dfrac{4fL_e}{D}$

where

f = Fanning friction factor
K = resistance coefficient of fitting
L_e = equivalent length, ft.
D = inside diameter of pipe.

$$\therefore\ L_e = \frac{KD}{4f}$$

(Omit the factor 4 if you use Moody-Darcy chart to get friction factor f'.)

$$\text{Now, velocity through the pipe} = \frac{1350\,\text{gal/min}}{60\,\text{s/min} \times 7.48\,\text{gal/ft}^3 \times 0.3474\,\text{ft}^2} = 8.64 \text{ ft/s}$$

$$\text{Reynolds number} = \frac{Du\rho}{\mu} = \frac{0.6651 \times 8.64 \times 74.9}{1.2 \times 0.000672} = 5.34 \times 10^5$$

From Fanning friction chart, $f = 0.0038$

$$\text{Then, the equivalent length of all fittings} = \frac{(\Sigma K)D}{4f} = \frac{8.22 \times 0.6651}{4 \times 0.0038} = 360 \text{ ft}$$

∴ total equivalent length of the discharge line = straight length + equivalent length of fittings = 400 + 360 = 760 ft

3.43 **a.** The levels in A and B are such that the line loss between A and D is much greater and so that the local pressure at D is the same from either A or B. Therefore, flow will be in the direction of C from both tanks A and B giving

$$Q_1 + Q_2 = Q_3$$

3.44 **b.** Head for adiabatic compression can be calculated by the following equation:

$$H_{ad} = \frac{k}{k-1}\frac{RT_1}{M}\left[\left(\frac{P_2}{P_1}\right)^{\frac{k-1}{k}} - 1\right] \text{ft}\cdot\text{lb}_f/\text{lb}$$

$$k = 1.4 \qquad P_2 = 35 + 14.7 = 49.7 \text{ psia}$$
$$P_1 = 0 + 14.7 = 14.7 \text{ psia}$$
$$M = 29.2$$
$$T_1 = 80 + 460 = 540°\text{R}$$

$$R = 10.731\,\frac{\text{psia}\cdot\text{ft}^3}{\text{lbmol}\cdot°\text{R}} \times \frac{144\,\text{in}^2}{\text{ft}^2} = 1545.3\,\frac{\text{psf}\cdot\text{ft}^3}{\text{lbmol}\cdot°\text{R}}$$

Therefore, substituting into the equation for H_{ad}

$$\therefore H_{ad} = \frac{1.4}{1.4-1}\frac{1545.3 \times 540}{29.2}\left[\left(\frac{49.7}{14.7}\right)^{\frac{0.4}{1.4}} - 1\right]$$
$$= (3.5)(28577.5)(0.4163)$$
$$= 41{,}638.9 \text{ ft}\cdot\text{lb}_f/\text{lb}$$

$$\text{Flow rate} = (1000/60) = 16.67 \text{ lb/min}$$

$$\text{Adiabatic horsepower} = \frac{41{,}638.9 \times 16.67}{33{,}000} = 21.03$$

3.45 **a.** For adiabatic compression, the theoretical discharge temperature is given by

$$\frac{T_2}{T_1} = \left(\frac{P_2}{P_1}\right)^{\frac{k-1}{k}} \qquad \therefore T_2 = T_1\left(\frac{P_2}{P_1}\right)^{\frac{k-1}{k}}$$

$$T_2 = 540\left(\frac{49.7}{14.7}\right)^{\frac{1.4-1}{1.4}} = 764.8°R = 304.8°\text{F}$$

3.46 **b.** With adiabatic efficiency taken into consideration, the discharge temperature is given by

$$T_2 = T_1 + \frac{T_1\left[\left(\frac{P_2}{P_1}\right)^{(k-1)/k} - 1\right]}{\varepsilon_{ad}}$$

$$= 540 + \frac{540\left[\left(\frac{49.7}{14.7}\right)^{(1.4-1)/1.4} - 1\right]}{0.8}$$

$$= 540 + 281 = 821°\text{R} = 361°\text{F}$$

3.47 **d.** If the compression is isothermal, the theoretical horsepower is given by

$$H_{iso} = \frac{RT_1}{M} \ln \frac{P_2}{P_1}$$

and $$\text{hp} = \frac{RT_1}{M}\frac{16.67}{33{,}000} \ln \frac{P_2}{P_1} = \frac{1545.3 \times 540}{29.2}\left(\frac{16.67}{33{,}000}\right) \ln \frac{49.7}{14.7} = 17.6$$

3.48 **d.** From the given data first calculate the average velocity of the flow through the pipe.

$$u_{av} = \frac{325}{7.48 \times 60} \times \frac{1}{0.0884} = 8.2 \text{ ft/s}$$

Calculate the Reynolds number.

$$\text{Re. No.} = \frac{d_i u \rho}{\mu} = \frac{0.3355 \times 8.2 \times 56.1}{470 \times 0.000672} = 488.7$$

The flow is streamline. Therefore, the average velocity in the pipe is 0.5 times the maximum velocity at the center of pipe. That is, $u_{av}/u_{max} = 0.5$

Then, $u_{max} = (u_{av}/0.5) = 8.2/0.5 = 16.2$ ft/s

Velocity at any point r between $r = 0$ and $r = r_w$ is given by the relation

$$u = u_{max}\left[1 - \left(\frac{r}{r_w}\right)^2\right]$$

where

r_w = distance from the center of pipe to the wall or the radius of pipe and

r = distance of a point from the center of pipe.

For a point midway between the center of the pipe and the wall, $\frac{r}{r_W} = 0.5$. The velocity at a point midway between the center of the pipe and the wall of the pipe is

$$u = 16.2\left[1 - (0.5)^2\right] = 16.2 \times 0.75 = 12.15 \text{ ft/s}$$

3.49 **d.** First calculate the flow area and the wetted perimeter. In the given figure, angle θ is given by cos $\theta = 6/r = 6/12 = 0.5$

Hence, $\theta = 60°$ and $\alpha = 30°$, $b = 6/\tan\alpha = 6/\tan 30° = 10.4"$

The flow area covered by triangles A and B $= 2\left\{\frac{1}{2} \times 10.4 \times 6\right\}$

$= 62.4$ in.^{2}

The flow area covered by C $= 0.7954(24^2)\ \dfrac{180 + 2 \times 30}{360} = 301.6$ in.^{2}

The total flow area = 62.4 + 301.6 = 364 in.^{2}

Now calculate the wetted perimeter.

$$\text{Wetted perimeter} = \pi(24)\left(\frac{240}{360}\right) = 52.27 \text{ in.}$$

$$\text{Equivalent diameter} = 4R_H = 4 \times \frac{\text{flow area}}{\text{wetted perimeter}}$$

$$= 4 \times \frac{364}{50.27} = 28.964" \approx 29"$$

3.50 **d.** The velocity of liquid flowing in a pipe is given by

$$u = \frac{Q}{A_C}$$

where
Q = volumetric flow rate and
A_C = flow or cross sectional area of pipe.

Thus, when A_C is constant, $u \propto Q$ Therefore, if flow is increased by 10%, the velocity will also increase by 10%. Next, the pressure drop for a flowing liquid in a pipe is given by the equation

$$\Delta P = \frac{2fLu^2\rho}{g_c D_i} \ \frac{\text{lb}_\text{f}}{\text{ft}^2}\left[\frac{\text{N}\cdot\text{m}}{\text{kg}}\right]$$

When the flow is increased, L, ρ, and g_c are the same. Since the flow was fully turbulent before increasing the same by 10%, it will be fully turbulent after the flow is increased.

Therefore, f will also be more or less constant. Then the pressure drops in conditions 1 and 2 will be given by

$$\frac{\Delta P_2}{\Delta P_1} = \left(\frac{u_2}{u_1}\right)^2 = \left(\frac{1.1}{1.0}\right)^2 = 1.21$$

Thus, if the flow is increased by 10%, the pressure drop will increase by 21%.

HEAT TRANSFER

3.51 c.

$$q = \frac{\Delta t}{R_T}$$

$$R_T = \frac{\Delta t}{q} = \frac{1500 - 180}{186.4} = \frac{1320}{186.4}$$

$$= 7.082 \text{ h}\cdot\text{ft}^2\cdot{}^\circ\text{F/Btu} \quad \text{(total resistance)}$$

$$\text{Resistance of fire brick} = \frac{6/12}{0.08(1 \text{ ft}^2)} = 6.25 \text{ h}\cdot\text{ft}^2\cdot{}^\circ\text{F/Btu}$$

Therefore, resistance of common brick = 7.082 – 6.25
= 0.832 h·ft²·°F/Btu

ΔX of common brick = (0.8)(1)(0.832) = 0.6656 ft = 7.9872" ≅ 8"

3.52 **b.** Temperature drop across fire brick = (6.25/7.082)(1320) = 1165 °F

Interface temperature between firebrick and common brick
= 1500 – 1165 = 335°F

3.53 **b.** Neglecting the metal wall resistance, relation for overall dirty coefficient of heat transfer based on the outside area can be written as follows:

$$\frac{1}{U_{do}} = \frac{1}{h_o} + \frac{1}{h_{do}} + \frac{1}{h_i}\frac{d_o}{d_i} + \frac{1}{h_{di}}\frac{d_o}{d_i}$$

$$\frac{1}{28.03} = \frac{1}{38.4} + \frac{1}{h_{do}} + \frac{1}{300} \times \frac{1}{0.902} + 0.003\frac{1}{0.902}$$

$$\frac{1}{h_{do}} = R_{do} = 0.035676 - 0.02604 - 0.0036955 - 0.003261$$

$$= 0.00268 \text{ h·ft}^2\text{·°F/Btu}$$

3.54 **a.**

$$t_f = \frac{220+100}{2} = 160\text{°F}$$

ρ_f = 61.28 lb/ft³ μ_f = 1.221 cP $g = 4.18 \times 10^8$ lb/h·ft
C_{pf} = 0.4751 Btu/lb·°F $\beta_f = 4.9 \times 10^{-4}$ °F^{-1}
k_f = 0.09379 Btu/h·ft²·°F/ft

$$h_c = 116\left[\frac{k_f^3\rho_f^2C_{pf}\beta_f}{\mu_f'}\left(\frac{\Delta t}{d_o}\right)\right]^{0.25}$$

$$= 116\left[\frac{0.09379^3 \times 61.28^2 \times 0.4751 \times 4.9 \times 10^{-4}}{1.221}\left(\frac{120}{1}\right)\right]^{0.25}$$

$$= 59.9 \text{ Btu/h·ft}^2\text{·°F}$$

3.55 **b.** OD of pipe = 2.38"
OD of pipe with insulation = 2.38 + 2 × 0.5 = 3.38"
Radiation area per foot of pipe = $\pi D_o L = \pi(3.38/12)(1) = 0.885$ ft²
T_s = 125 + 460 = 585°R T_a = 70 + 460 = 530°R

$$q = A\varepsilon(0.1713)\left[\left(\frac{T_S}{100}\right)^4 - \left(\frac{T_a}{100}\right)^4\right]$$

$$= 0.885(0.9)(0.1713)[(5.85)^4 - (5.3)^4]$$

$$= 52.1 \text{ Btu/h · ft}^2 \text{ · linear ft of pipe}$$

$$h_r = \frac{q}{A(T_S - T_a)} = \frac{52.1}{0.885(585-530)} = 1.08 \text{ Btu/hr · ft}^2 \text{ · °F}$$

$$h_c = 116\left[\frac{k_f^3\rho_f^2 C_{pf}\beta_f}{\mu_f'}\left(\frac{\Delta t}{d_o}\right)\right]^{0.25}$$

$$= 116\left[\frac{0.09379^3 \times 61.28^2 \times 0.4751 \times 4.9 \times 10^{-4}}{1.221}\left(\frac{120}{1}\right)\right]^{0.25}$$

$$= 59.9 \text{ Btu/h}\cdot\text{ft}^2\cdot{}^\circ\text{F}$$

3.56 c. $t_b = \dfrac{80+100}{2} = 90°\text{F},\quad \mu_b = 18 \text{ cP},\quad D_i = 0.902 \text{ in.} = 0.0752 \text{ ft}$

$N_{Re} = \dfrac{0.0752 \times 4.5 \times 56}{18 \times 0.000672} = 1567$ This is streamline flow.

$$\frac{h_i D_i}{k_b} = 1.86\left[\left(\frac{D_i G}{\mu_b}\right)\left(\frac{C_{Pf}\mu_b}{k_b}\frac{D_i}{L}\right)\right]^{1/3}\left(\frac{\mu_b}{\mu_w}\right)^{0.14}$$

$$\frac{D_i G}{\mu_w} = 1567$$

$$\frac{C_P\mu_b}{k_b} = \frac{0.48 \times 18 \times 2.42}{0.08} = 261.4$$

$$\frac{\mu_b}{\mu_w} = \frac{18}{3.6} = 5 \qquad \frac{k_b}{D_i} = \frac{0.08}{0.0752} = 1.0638$$

$$\frac{D_i}{L} = \frac{0.0752}{12} = 0.00627$$

$$h_c = 1.86 \times (1.0638)[1567 \times 261.4 \times 0.00627]^{1/3}(5)^{0.14}$$

$$= 34.1 \text{ Btu/h}\cdot\text{ft}^2\cdot{}^\circ\text{F}$$

3.57 b. Number of tubes per pass = 172/4 = 43 tubes

Cross sectional area = 0.546 in.^{2} × 43/144 = 0.163 ft^2

Mass velocity through tubes G = 154,000/0.163 = 944,785 lb/h·ft^2

The Dittus-Boelter equation uses properties at bulk temperature of the fluid.

$$t_b = \frac{80+120}{2} = 100°\text{F} \qquad D_i = 0.834/12 = 0.0695 \text{ ft}$$

$$\mu = 1.66 \text{ lb/h}\cdot\text{ft}, \qquad k = 0.363 \text{ Btu/h}\cdot\text{ft}^2\cdot{}^\circ\text{F/ft}$$

$$\left(\frac{D_i G}{\mu}\right)^{0.8} = \left(\frac{0.0695 \times 944,785}{1.66}\right)^{0.8} = 4762$$

$$\left(\frac{C_P\mu}{k}\right)^{0.4} = \left(\frac{1 \times 1.66}{0.363}\right)^{0.4} = 1.837$$ since water is being heated.

The Dittus-Boelter equation is

$$\frac{h_i D_i}{k} = 0.023\left(\frac{D_i G}{\mu}\right)^{0.8}\left(\frac{C_P \mu}{k}\right)^{0.4} \quad \text{for heating of fluid.}$$

Therefore,
$$h_i = 0.023\left(\frac{k}{D_i}\right)\left(\frac{D_i G}{\mu}\right)^{0.8}\left(\frac{C_P \mu}{k}\right)^{0.4}$$

$$= 0.023\left(\frac{0.363}{0.0695}\right)(4762)(1.837)$$

$$= 1050 \text{ Btu/h}\cdot\text{ft}^2\cdot\text{°F}$$

3.58 **a.** The shell side flow area is given by

$$a_s = \frac{ID \times C \times B}{P_T \times 144} = \frac{21.25 \times 0.25 \times 6}{1.25 \times 144} = 0.1771 \text{ ft}^2$$

$$G_S = \frac{100{,}000}{0.1771} = 564{,}706 \text{ lb/h}\cdot\text{ft}^2$$

3.59 **c.** The shell side heat transfer coefficient is given by

$$\frac{h_o D_e}{k} = 0.36\left(\frac{D_e G_S}{\mu}\right)^{0.55}\left(\frac{C_p \mu}{k}\right)^{\frac{1}{3}}\left(\frac{\mu}{\mu_w}\right)^{0.14}$$

The equivalent diameter on the shell side, d_e, for a triangular pitch is

$$d_e = \frac{4\left(\frac{1}{2}P_T \times 0.86 P_T - \frac{1}{2}\frac{\pi d_o^2}{4}\right)}{\frac{1}{2}\pi d_o}$$

$$= \frac{4(0.5 \times 1.25 \times 0.86 \times 1.25 - 0.5 \times \pi \times 1^2/4)}{0.5 \times \pi \times 1}$$

$$= 0.7109 \text{ in.} = 0.0592 \text{ ft}$$

$$t_B = \frac{190 + 120}{2} = 155\text{°F}$$

$$\mu = 0.76 \text{ cP} = 0.76(2.42) = 1.84 \text{ lb/h}\cdot\text{ft}$$

$$\left(\frac{D_e G_S}{\mu}\right)^{0.55} = \left(\frac{0.592 \times 564{,}706}{1.84}\right)^{0.55} = 220$$

$$\left(\frac{C_p \mu}{k}\right)^{\frac{1}{3}} = \left(\frac{0.88 \times 1.84}{0.34}\right)^{\frac{1}{3}} = 1.68$$

Because of low viscosity value, assume $\mu/\mu_w = 1$.

Then, $h_o = 0.36(0.34/0.0592)(220)(1.68) = 769$ Btu/h·ft^2·°F

3.60 **b.** Log mean diameter $= \dfrac{1-0.902}{\ln(1/0.902)} = 0.95016$ in.

$$\frac{1}{U_C} = \frac{1}{h_o} + \frac{1}{h_i}\frac{d_o}{d_i} + \frac{l_w}{k_w}\frac{d_o}{d_{av}}$$

$$= \frac{1}{250} + \frac{1}{700}\frac{1}{0.902} + \frac{0.049/12}{26}\frac{1}{0.95016} = 0.005749$$

Therefore, $U_C = 173.94 \approx 174$ Btu/h·ft²·°F

$$\frac{1}{U_D} = \frac{1}{U_C} + \frac{1}{h_{do}} + \frac{1}{h_{di}}\frac{d_o}{d_i} = \frac{1}{174} + 0.0015 + 0.001\frac{1}{0.902}$$

$$= 0.008357647$$

$$U_D = 119.65 \text{ Btu/h·ft}^2\text{·°F}$$

3.61 **d.** From the Wilson plot, the equations for the lines are

$$\frac{1}{U_0} = \frac{1}{C}\frac{1}{u^{0.8}} + 0.00092 \quad \text{for the dirty tube}$$

and

$$\frac{1}{U_0} = \frac{1}{C}\frac{1}{u^{0.8}} + 0.0004 \quad \text{for the clean tube}$$

The intercept in the case of the dirty tube consists of three resistances:

Intercept (dirty tube) $= R_v + R_w + R_d$
Intercept (clean tube) $= R_v + R_w$

where
R_v = Resistance of steam film, h·ft²·°F/Btu
R_w = Resistance of wall, h·ft²·°F/Btu
R_d = Resistance due to fouling, h·ft²·°F/Btu

At a velocity of 1 ft/s, the reciprocal of the slope of either line gives the valuc of the inside film coefficient based on the outside area of tube.

Hence, since the slope of the lines is $1/C = 0.0038$

$$h_{io} = 1/0.0038 = 263.2 \text{ Btu/h·ft}^2\text{·°F}$$

and h_i based on inside area = 263.2(1/0.902) = 291.8 Btu/h·ft²·°F

3.62 **a.** The intercept of the line for the clean tube $= R_v + R_w = 0.0004$

Tube wall resistance = 0.000068

Then, $R_v = 0.0004 - 0.000068 = 0.000332$

Steam film coefficient = 1/0.000332 = 3012 Btu/h·ft²·°F

3.63 **d.** Intercept of the line for the dirty tube $= R_v + R_w + R_d$
= 0.00092 from the Wilson plot

Intercept of the line for the clean tube $= R_v + R_w = 0.0004$

Therefore, $R_{do} = 0.00092 - 0.0004 = 0.00052$

Then $h_{do} = 1/0.00052 = 1923$ Btu/h·ft²·°F based on outside area

$h_{di} = 1923(1/0.902) = 2132$ Btu/h·ft²·°F

3.64 **a.** Resistance of the two insulation layers

$$\frac{1.25/12}{0.058\left(\pi\frac{3.48}{12}\right)}+\frac{2.5/12}{0.042\left(\pi\frac{7.07}{12}\right)} = 1.9713 + 2.68 = 4.651 \text{ h·ft}^2\text{·°F/Btu}$$

3.65 **c.** Resistance of the pipe wall

$$R_w = \frac{0.154/12}{26\left(\pi\frac{2.22}{12}\right)} = 0.00085 \text{ h·ft}^2\text{·°F/Btu}$$

$$\text{Heat loss per foot of pipe} = q = \frac{900-120}{0.00085+1.971+2.68}$$
$$= 167.7 \text{ Btu/h·ft of pipe.}$$

3.66 **a.** The temperature drop is directly proportional to the resistance.

Resistance up to the interface of the insulation layers = 0.00085 + 1.971 = 1.97185

Total resistance up to the surface of the second layer = 0.00085 + 1.971 + 2.68 = 4.65185

The temperature drop up to the interface of the insulation layers

$$= \frac{1.97185}{4.65185} \times (900 - 120) = 330.6\text{°F} \approx 331\text{°F}$$

The interface temperature between the two insulation layers = 900 – 331 = 569°F

3.67 **b.** The surface coefficient of heat transfer,

$$h_a = \frac{q}{A_o \Delta t} = \frac{167.7}{\pi\left(\frac{9.875}{12}\right)(1)(120-85)} = 1.85 \text{ Btu/ h·ft}^2\text{·°F}$$

$$h_a = h_c + h_r$$

$$h_c = 0.5\left(\frac{\Delta t}{d_o}\right)^{0.25} = 0.5\left(\frac{120-85}{9.875}\right)^{0.25} = 0.69 \text{ Btu/ h·ft}^2\text{·°F}$$ from the insulated surface

Therefore, $h_r = h_a - h_c = 1.85 - 0.69 = 1.16$ Btu/ h·ft²·°F from the insulated surface

3.68 **c.**

$$A_1 = \pi\left(\frac{2.38}{12}\right)(1) = 0.622 \text{ ft}^2$$

$$A_2 = 4(1)(1) = 4 \text{ ft}^2$$

In this case, $F_A = 1$, $F_e = \dfrac{1}{\frac{1}{\epsilon_1} + \frac{A_1}{A_2}\left(\frac{1}{\epsilon_1} - 1\right) - 1} = \dfrac{1}{\frac{1}{0.8} + \frac{0.622}{4}\left(\frac{1}{0.28} - 1\right)}$

$$= 0.606$$

$$Q = F_A F_e A(\sigma)\left(T_p^4 - T_d^4\right) = (1)(0.622)(0.1713 \times 10^{-8})(760^4 - 530^4)$$
$$= 164.4 \text{ Btu/h·ft of pipe}$$

3.69 **c.** The surroundings are very large. Therefore, radiation reflected back to the pipe can be neglected. Also, assume negligible resistance from steam film and metal.

Surface temperature, $t_s = 325 + 460 = 785°R$

Air temperature, $t_s = 70 + 460 = 530°R$

$$\text{Heat loss from bare pipe} = 0.79(0.173 \times 10^{-8})(785^4 - 530^4)$$
$$= 411 \text{ Btu/h·ft}^2$$
$$\text{Heat loss from painted pipe} = 0.35(0.173 \times 10^{-8})(785^4 - 530^4)$$
$$= 182.1 \text{ Btu/h·ft}^2$$

$$\% \text{ decrase in radiation loss} = \frac{411 - 182}{411} \times 100 = 55.7\%$$

3.70 **d.** This problem can be solved by the trial and error method, but it can also be solved directly with the use of the effectiveness factor as follows:

$$C_H = 45{,}300 \times 0.452 = 20{,}746 \quad C_C = 10{,}600 \times 1 = 10{,}600$$

$$\frac{C_{\min}}{C_{\max}} = \frac{10{,}600}{20{,}476} = 0.518$$

$$\frac{UA}{C_{\min}} = \frac{53 \times 325}{10{,}600} = 1.625$$

Effectiveness factor $\varepsilon = \dfrac{1 - e^{-1.625(1-0.518)}}{1 - 0.518\, e^{-1.625(1-0.518)}} = 0.711$

$$q = \varepsilon C_{\min}(T_{H1} - t_{c1}) = 0.711 \times 10{,}600(230 - 80) = 1{,}130{,}490 \text{ Btu/h}$$

Then $45{,}300(0.452)(T_{Hi} - T_{Ho}) = 1{,}130{,}490$

$$T_{Hi} - T_{Ho} = 55.2$$

$$T_{Ho} = 230 - 55.2 = 174.8 \approx 175°F$$

MASS TRANSFER

3.71 **a.** Mole fractions of non-diffusing component water are

$$x_{B1} = 1 - 0.0323 = 0.9677$$

$$x_{B2} = 1 - 0.0092 = 0.9908$$

$$x_{BM} = \frac{0.9908 - 0.9677}{\ln(0.9908/0.9677)} = 0.9792$$

$$\left(\frac{\rho}{M}\right)_{av} = \frac{3.26 + 398}{2} = 3.329 \frac{\text{lb mol}}{\text{ft}^2 \cdot \text{h}}$$

Using the equation for diffusion of A in non-diffusing B,

$$N_A = \frac{D_{AB}}{z x_{BM}} \left(\frac{\rho}{M}\right)_{av} (x_{A1} - x_{A2})$$

$$= \frac{4.8 \times 10^{-5} \frac{\text{ft}^2}{\text{h}}}{0.00328 \text{ ft} \times (0.9792)} \left(\frac{3.329 \text{ lb mol}}{\text{ft}^3}\right)(0.0323 - 0.0092)$$

$$= 1.15 \times 10^{-3} \text{ lb mol/ft}^2 \cdot \text{h}$$

3.72 b.

$$D_{AB} \propto \frac{T^{1.75}}{P}$$

$T_2 = 273 + 60 = 333$ K, $T_1 = 298$ K $P_1 = 1$ atm, $P_2 = 2$ atm

Then

$$(D_{AB})_{T_2} = \left(\frac{T_2}{T_1}\right)^{1.75} \left(\frac{P_1}{P_2}\right) (D_{AB})_{T_1}$$

$$= \left(\frac{333}{298}\right)^{1.75} \left(\frac{1}{2}\right) \times 0.96 \times 10^{-5}$$

$$= 0.583 \times 10^{-5} \text{ cm}^2\text{/s}$$

Now convert to units desired

$$D_{AB} = 0.583 \times 10^{-5} \frac{\text{cm}^2}{\text{s}} \times \left(\frac{1}{30.48^2}\right) \frac{\text{ft}^2}{\text{cm}^2} \times \frac{3600 \text{ s}}{\text{h}} = 2.26 \times 10^{-5} \frac{\text{ft}^2}{\text{h}}$$

3.73 b.

$$z = \frac{50 - 25}{2} \times \frac{1}{1000} = 0.125 \text{ m}$$

Mean area of diffusion for cylindrical shape,

$$S_{av} = \frac{2\pi L(r_o - r_i)}{1000 \ln \frac{r_o}{r_i}} = \frac{2\pi(1)(50 - 25)}{2 \times 1000 \ln\left(\frac{50}{25}\right)} = 0.1113 \text{ m}^2$$

Solubility of H_2 in neoprene at 2 atm = 0.051 = 0.051 m^3(STP)/m^3 solid

Therefore, concentration at inner surface,

$$c_{A1} = \frac{0.051 \times 2}{22.414} = 4.55 \times 10^{-3} \frac{\text{kg mol}}{\text{m}^3\text{solid}}$$

$$c_{A2} = 0$$

Then $$N_A = \frac{D_{AB}(c_{A1} - c_{A2})}{z_2 - z_1}$$

$$= \frac{(1.03\times10^{-10})(0.1113)\left[4.55\times10^{-3} - 0\right]}{0.0125}$$

$= 4.173 \times 10^{-12}$ kg mol/s per meter length of pipe.

3.74 **b.** For an ideal binary system,

$$P_A x_A + P_B x_B = P$$

$$P_A x_A + P_B(1 - x_A) = P \qquad \text{(a)}$$

$$y_A = \frac{p_A}{P} = \frac{P_A x_A}{P} \qquad \text{(b)}$$

From equation (a), $197.3x_A + 37.1(1 - x_A) = 101.32$

Solving the equation, $$x_A = \frac{101.32 - 37.1}{197.3 - 37.1} = 0.4009$$

Then $$y_A = \frac{197.3\times0.4009}{101.32} = 0.781$$

3.75 **b.** On the *t*-*x* diagram, the azeotropic point is shown as E. Reading the composition, the water mole fraction in the azeotrope is 0.67. Therefore, the mole fraction of i-butanol = 1 – 0.67 = 0.33

$$\text{wt \% of i-butanol in azeotrope} = \frac{0.33(71.14)}{0.33(71.14) + 0.67(18.02)} \times 100$$

$$= 66.04 \text{ wt \%}$$

3.76 **c.** From *t*-*x* diagram, at 80°C, the composition of butanol-rich layer is

mole fraction of water = 0.56

mole fraction of butanol = 1 – 0.56 = 0.44

$$\text{Therefore, wt \% of i-butanol} = \frac{0.44(71.14)}{0.44(71.44) + 0.56(18.02)}$$

$$= 75.62 \text{ wt \%}$$

3.77 **c.** mole fraction of water in vapor = 1 – 0.15 = 0.85

Locate point ($y_w = 0.85$) and $t = 110$ on the equilibrium diagram, and move down vertically to meet *y* vs. *t* curve. At the intersection point, read temperature to the right or left. The temperature is 92.8°C, approximately.

3.78 **b.** Using the given equilibrium diagram, the following points are located on it. The point, *D*, denotes the composition of heptane in distillate. Since feed is saturated liquid, the *q* line is vertical, and it is drawn from $x_F = 0.4$ to meet the equilibrium curve at *M*. Joining *D* and *M* gives the operating line for minimum reflux. The construction is shown in Exhibit 3.78.

minimum reflux operating line intersects the y axis at $y = 0.45$

$$\frac{x_D}{R_M + 1} = 0.45 \qquad R_m = \frac{0.97}{0.45} - 1 = 1.156$$

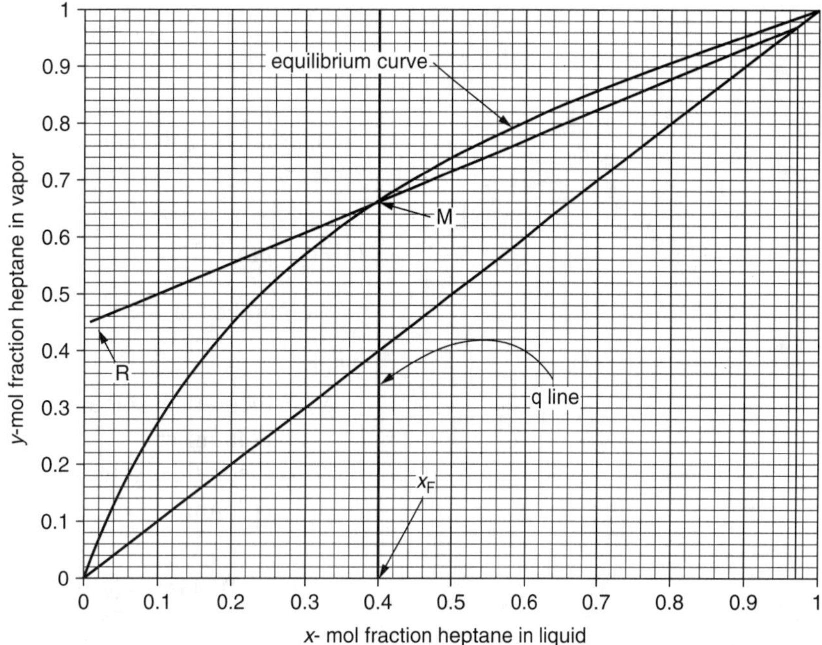

Exhibit 3.78 Construction of minimum reflux line, Problem 3.78

3.79 **a.** Key components are those between which the separation is specified and desired in a distillation column. In this problem separation between phenol and o-cresol is specified. Phenol is light key because it is recovered in distillate and o-cresol is heavy key as it is recovered in bottoms.

3.80 **b.** The minimum number of theoretical stages by the Fenske equation

$$N_m = \frac{\log\left[\left(\frac{x_{LD}}{x_{HD}}\right)\left(\frac{x_{HB}}{x_{LB}}\right)\right]}{\log(\alpha_{AV})} = \frac{\left[\left(\frac{0.953}{0.0455}\right)\left(\frac{0.202}{0.0524}\right)\right]}{\log(1.25)} = 19.7$$

3.81 **c.** $R_m = 5.3622$ (given) $\qquad R = 10$ (given)

$$\frac{R}{R+1} = \frac{10}{10+1} = 0.91$$

$$\frac{R_m}{R_m+1} = \frac{5.3622}{5.3622+1} = 0.843$$

Using the Erbar-Maddox correlation,

for $\frac{R}{R+1} = 0.91$ and parameter $\frac{R_m}{R_m+1} = 0.843$,

one gets the ratio $N_M/N \cong 0.75$

Thus, $\qquad N = N_M/0.75 = 19.7/0.75 = 26.3$ theoretical stages

3.82 **a.** First, determine the feed plate location. The Kirkbride equation is

$$\log\left(\frac{N_R}{N_S}\right) = 0.206\log\left[\left(\frac{x_{HF}}{x_{LF}}\right)\left(\frac{B}{D}\right)\left(\frac{x_{LB}}{x_{HD}}\right)^2\right]$$

$$= 0.206\log\left[\left(\frac{0.15}{0.35}\right)\left(\frac{66.95}{33.05}\right)\left(\frac{0.0524}{0.0455}\right)^2\right]$$

Completing the calculation gives: $\frac{N_R}{N_S} = 1.0295$

$\therefore\ N_R = 1.0295\ N_s$

In Problem 3.81,

the number of theoretical stages $= N_R + N_s = 26.3$

where

N_R = theoretical stages in the rectification section of the distillation column and

N_s = theoretical stages in the stripping section of the distillation column

$N_R = 1.0295N_s + N_s = 26.3$ and thus, $N_s = (26.3/2.0295) = 12.96$

$N_R = 26.3 - 12.96 = 13.34$ theoretical stages counting from top

3.83 **b.** Average molecular weight = 0.05(17) + 29(0.95) = 28.4

$$\rho_g = \frac{28.4}{359}\frac{492}{528}\frac{15}{14.7} = 0.0752\ \text{lb/ft}^3$$

Gas flow rate = 30,000(0.0752) = 2256 lb/h

$$\frac{L}{G}\sqrt{\frac{\rho_g}{\rho_L}} = 1\sqrt{\frac{0.0752}{62.4}} = 0.0347 \qquad \frac{\rho_w}{\rho_L} = \frac{62.4}{50} = 1.248$$

For $\Delta P = 0.5"$ H_2O, Y (ordinate) = 0.058 from the pressure drop chart. At flood point and abscissa = 0.0347, $Y = 0.19$

$$\left(\frac{G_L}{G_f}\right)^2 = \frac{0.058}{0.19} \qquad \frac{G_L}{G_f} = \sqrt{\frac{0.058}{0.19}} = 0.55$$

Percent flood = 0.55(100) = 55%

3.84 **c.**

$$\frac{L}{G}\sqrt{\frac{\rho_G}{\rho_L}} = \frac{121{,}500}{79{,}000}\sqrt{\frac{0.288}{46.4}} = 0.12$$

From Exhibit 3.84 (provided with the problem), for $X = 0.12$ and tray spacing of 24", Y ordinate = 0.32.

$$u_f = 0.32\left(\frac{\sigma}{20}\right)^{0.2}\left(\frac{\rho_L - \rho_g}{\rho_g}\right)^{0.5} = 0.32\left(\frac{18}{20}\right)^{0.2}\left(\frac{46.4 - 0.288}{0.288}\right)^{0.5}$$

$$= 3.97\ \text{ft/s}$$

Use u_a = 80% of flood.

$$u_a = 0.8(3.97) = 3.18 \text{ ft/s actual velocity.}$$

Active area, $$A_a = \frac{76.13}{3.18} = 23.94 \text{ ft}^2$$

$$A_t = A_a + 2A_d + A_w = 23.94 + 2(0.1)A_t + 0.075A_t$$

Total area, $$A_t = \frac{23.94}{0.725} = 33.02 \text{ ft}^2.$$

$$\text{Diameter of column} = \sqrt{\frac{33.02}{0.7854}} = 6.484 \text{ ft} \approx 6.5 \text{ ft}$$

3.85 **c.** By material balance,

$$Fx_F = Lx_L + V \cdot y_V$$

$$y_V = 0,\ F = 30{,}000 \text{ lb/h},\quad x_F = 0.1 \text{ mass fraction,}$$
$$x_L = 0.5 \text{ mass fraction}$$

Substitution gives 30,000(0.1) = L(0.5)

$$L = 30{,}000 \times 0.1/0.5 = 6000 \text{ lb/h}$$

Total evaporation = 30,000 – 6000 = 24,000 lb/h

Saturation temperature of water at 4" Hg = 1.96 psia = 125°F

Enthalpy of superheated vapor = 1116 + 0.46(198 – 125)
= 1149.6 Btu/lb

By heat balance, 30,000 × 60 + S(1164 – 218.5) = 24,000(1149.6)
+ 6000(222)

Solution of the above equation gives S = 28671 lb/h
Steam economy = 24,000/28,671 = 0.837 lb evaporation/lb of live steam

3.86 **c.** From steam tables, h_l = 54.03 Btu/lb at 85°F, h_l = 87.97 Btu/lb at 120°F

H_V = 1149.6 Btu/lb (refer to Problem 3.85 solution)

By heat balance on the condenser,

$$24{,}000(1149.6) + W(54.03) = (W + 24{,}000)(87.97)$$

From which, W = 750,048 lb/h
= 750,048/500 = 1500 gpm

3.87 **c.** Make enthalpy and moisture balances. The operation can be represented as follows:

$$H_1 = 0.24(110) + Y_1[1061 + 0.45(110)] = 26.4 + 1110.5Y_1 \text{ Btu/lb dry air}$$

Enthalpy balance on the dryer gives (assume M = mass of dry air lb/h)

$$71.9M + 14.4 \times 1700 + 48.02 \times 300 = M(26.4 + 1110.5Y_1)$$
$$+ 8 \times 107.88 + 1700(32.4)$$

which simplifies to $M(45.5 - 1110.5Y_1) = 17{,}057.04$

Moisture balance: $M(Y_1 - 0.01) = 1700(0.1765 - 0.005)$

Dividing one equation by the other gives

$$\frac{45.5 - 1110.5Y_1}{Y_1 - 0.01} = \frac{17,057.04}{1700 \times 0.1705} = 58.505$$

$$Y_1 = \frac{45.5 + 0.5805}{1164.505} = 0.0396 \text{ lb/lb dry air}$$

$$M = \frac{1700 \times 0.1715}{0.0396 - 0.01} = 9850 \text{ lb/h dry air}$$

3.88 d.

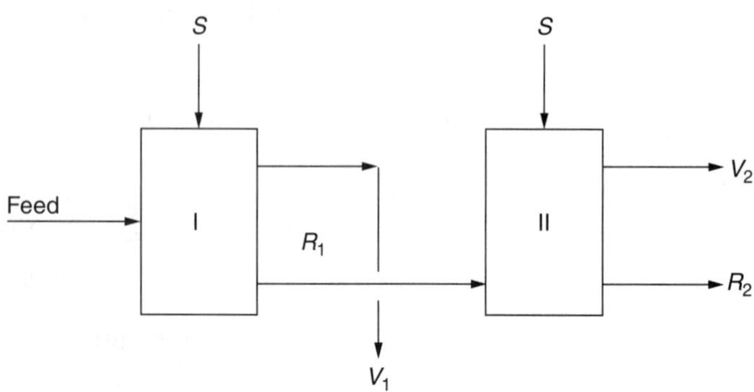

Exhibit 3.88a Cross current two-stage extraction for Problem 3.88

Total mixture of fresh solids and solvent = 3000 + 1000 = 4000 lb/h

Mass fraction solvent in the mixture = 3000/4000 = 0.75

Mass fraction solute in mixture = 500/4000 = 0.125

$R_1 = 500 + 500(1.2) = 1100$ lb/h

$E_1 = 4000 - 1100 = 2900$ lb/h

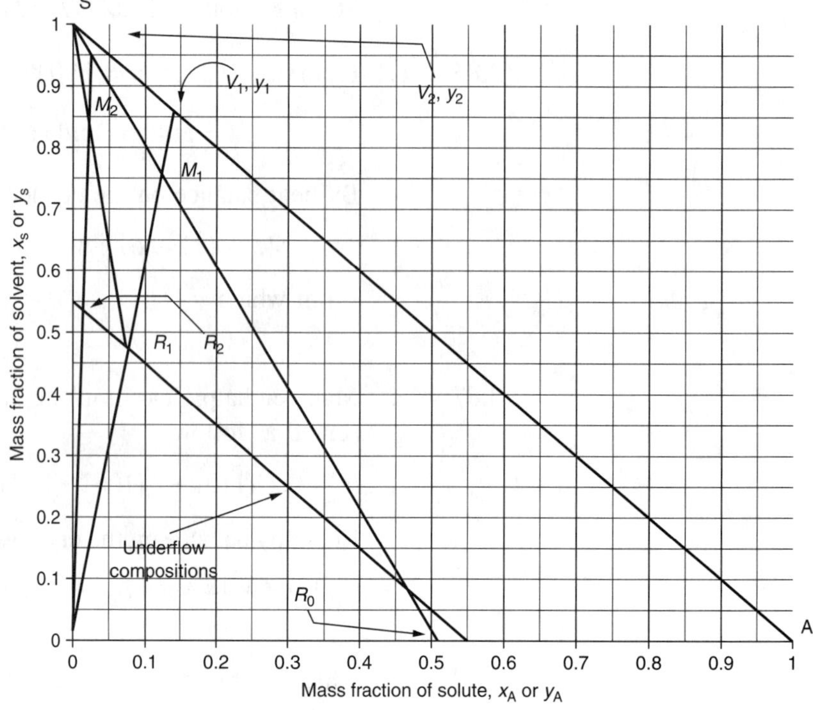

Exhibit 3.88b Leaching on triangular diagram, Problem 3.88

See Exhibit 3.88b for the solution. Locate R_0 on the x-axis at $x_A = 0.5$. Join R_0 and S and locate point M_1 representing the mixture with coordinates $x_s = 0.75$ and $x_A = 0.125$. Join M_1 and the origin to intersect the underflow curve in R_1 and the hypotenuse in $y_{i,}$ V_1.

From the diagram, $y_1 = 0.86$, $R_1 + S = 1100 + 3000 = 4100 = M_2$

The solvent in $M_2 = 3000 + 600(0.86) = 3516$

Join SR_1. Locate M_2 at $x_s = 0.86$. Join the origin and M_2 to meet the hypotenuse in V_2, y_2

$y_2 = 0.98$ $\quad V_2 = 4100 - 1100 = 3000$ lb/h

The solute recovered in two stages $= 2900(0.14) + 3000(0.02)$
$= 466$ lb/h

$$\text{Recovery} = \frac{466}{500} \times 100 = 93.2\%$$

3.89 **b.** Use the given phase diagram for solving the problem.

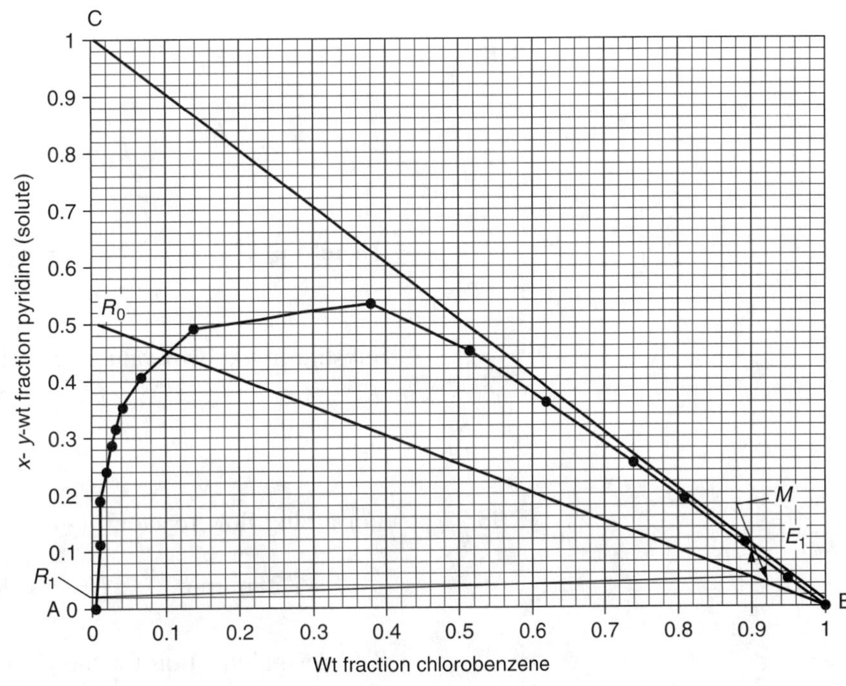

Exhibit 3.89 Liquid-liquid extraction on triangular diagram

The concentration in the raffinate (water layer) is to be reduced to 2%. This point must lie on the distribution curve and at $y = 0.02$. Utilizing this fact, the construction is made as shown in Exhibit 3.89.

The point R_0 (original pyridine-water solution) is connected with point B representing the solvent. Point R_1 is located on the distribution curve and the tie line passing through it is drawn to intersect R_0B at M, the mixture point, and to meet the distribution curve at E_1. The amount of solvent used is found from the line segment lengths R_0M and MB.

$$\text{Solvent used} = \frac{R_0M}{MB} = \frac{82}{7.5} \times 2000 = 21{,}867 \text{ lb/batch}$$

3.90 **c.** From the humidity chart for air with 85°F dry bulb and 75°F wet bulb temperature, the humidity is nearest to 0.0168 lb of water vapor per lb of dry air.

CHEMICAL REACTION ENGINEERING

3.91 **b.** $V = 3\ \text{m}^3, \quad C_{A0} = C_{B0} = 5\ \text{mol/liter} = 5\ \text{kg mol/m}^3$

$-r_A = k_1 = 1.405\ \text{kg mol/m}^3\cdot\text{min}$ (because the reaction rate is independent of concentrations)

$v_0 = 1.2\ \text{m}^3/\text{min}, \quad T = 300\ \text{K}, \quad E = 53.2\ \text{kJ/mol}$

For a continuous-flow reactor,

$(-r_A)V_{\text{cstr}} = v_0\,[C_{A0} - C_{A0}(1 - X_A)]$ by a material balance

By simplification and rearrangement,

$$X_A = \frac{(-r_A)V_{\text{cstr}}}{v_0 C_{A0}} = \frac{1.405(3)}{1.2(5)} = 0.7025$$

3.92 **a.** $T = 310\ \text{K}, \quad E = 53.2\ \text{kJ/mol}$

$$k_1 = \alpha e^{-\frac{E}{RT_1}} \qquad k_2 = e^{-\frac{E}{RT_2}}$$

Therefore,

$$k_2 = k_1 e^{-\frac{E}{R}\left(\frac{1}{T_2}-\frac{1}{T_1}\right)}$$

Substituting known values, $k_2 = 1.405\ e^{-\frac{53.2\times10^3}{8.314}\left(\frac{1}{310}-\frac{1}{300}\right)}$

$\cong 2.8\ \text{kg mol/m}^3\cdot\text{min}$

$= 2.8\ \text{g mol/liter}\cdot\text{min}$

3.93 **c.** For a plug flow reactor,

$v_0 = 1.2\ \text{m}^3/\text{min}, \quad T = 310\ \text{K}, \quad k_2 = 2.8\ \text{kg mol/m}^3\cdot\text{min}$

The design equation for the plug flow reactor is

$$\frac{V_{\text{plug flow}}}{v_0} = C_{A0}\int_0^{X_A}\frac{dX_A}{k_2} = C_{A0}\frac{X_A}{k_2}$$

Therefore,

$$V_{\text{plug flow}} = \frac{C_{A0}X_A}{k_2}\times v_0$$

Substituting known values,

$$V_{\text{plug flow}} = \frac{5(0.7025)}{2.8}\times 1.2 = 1.51\ \text{m}^3$$

3.94 **a.**

$$\Delta G^{\circ}_{298} = \sum\left(\Delta G^{\circ}_{298}\right)_{\text{products}} - \sum\left(\Delta G^{\circ}_{298}\right)_{\text{reactants}}$$
$$= 2(16.28) + 1(0) - 1(-3.75)$$
$$= 36.31 \text{ kcal/g mol}$$

3.95 **b.** Heat of reaction at 298 K = 2(12.5) + 1(0) – (–29.81) = 54.81 kcal/g mol

Net $\Delta C^{\circ}_p = 2(0.0028 + 3.00 \times 10^{-5}T) + (0.0069 + 0.4 \times 10^{-5}T)$

$$- (0.01178 + 4.268 \times 10^{-5}T)$$
$$= 0.00072 + 2.132 \times 10^{-5}T$$

$$I_H = \Delta H^{\circ}_{298} - \Delta\alpha T - \frac{1}{2}\Delta\beta T^2$$
$$= 54.81 - 0.00072(298) - 1.066 \times 10^{-5}(298)^2 = 53.65$$

$$T = 500 + 273 = 773 \text{ K}$$

$$\Delta H^{\circ}_{773} = 53.65 + 0.00072(773) + 1.066 \times 10^{-5}(773)^2$$
$$= 60.57 \text{ kcal/g mol}$$

3.96 **c.** To calculate K:

$$T = 273 + 704 = 977 \text{ K}$$

$$\Delta G^{\circ}_T = I_H + I_G T - \Delta\alpha T \ln T - \frac{1}{2}\Delta\beta T^2$$

$$36.31 = 53.65 + I_G\, 298 - 0.00072 \times 298 \ln 298 - 1.066 \times 10^{-5}(298^2)$$

From which $I_G = (36.31 - 53.65)/298 = -0.0509$

$$\Delta G^{\circ}_{977} = 53.65 - 0.0509(977) - 0.00072(977)\ln(977)$$
$$- 1.066 \times 10^{-5}(977^2) = -11.097 \text{ kcal/g mol}$$

$$\Delta G^{0}_{977} = -RT \ln K_{977}$$

$$\ln K_{977} = -\frac{-11.097 \times 10^3}{1.987 \times 977} = 5.716$$

$$K_{977} = e^{5.716} = 303.7$$

3.97 **a.** If X is the conversion,

n-butane remaining = $1 - X$
C_2H_4 produced = $2X$
H_2 produced = X
Total = $1 + 2X$

$$\text{In this case, } K = \frac{\left[\left(\frac{2X}{1+2X}\right)P_T\right]^2\left[\left(\frac{X}{1+2X}\right)P_T\right]}{\frac{1-X}{1+2X}P_T} = \frac{\left[\frac{4X^2}{(1+2X)^2}\right]\left(\frac{X}{1+2X}\right)P_T^3}{\frac{1-X}{1+2X}P_T}$$

$$= \frac{4X^3}{(1-X)(1+2X)^2}\frac{2^3}{2} \quad \text{since } P_T = 2 \text{ atm}$$

$$= \frac{16X^3}{(1-X)(1+2X)^2} = 303.7 \text{ (from Problem 3.96)}$$

This simplifies to $X^3 - 0.74025X - 0.24675 = 0$

This is a cubic equation. An acceptable root was found by trial and error numerically using a spreadsheet program. This gave $X = 0.9942$.

The cubic equation can be solved directly without trial and error by method of substitution as explained in Perry's handbook p. 2-15, 6th ed.

In the present case, the coefficient b is zero. Hence,

$$p = -0.74025 \quad \text{and} \quad q = -0.24675$$

Then
$$R = \left(\frac{p}{3}\right)^3 + \left(\frac{q}{2}\right)^2 = 0.000197878$$

Because b, c, and d are real and $R > 0$, the equation has one real root and two conjugate complex roots. The real root is given by $A + B$ where

$$A = \sqrt[3]{-q/2 + \sqrt{R}} \quad \text{and} \quad B = \sqrt[3]{-q/2 - \sqrt{R}}$$

Using values of q and R, $A = \sqrt[3]{-\frac{-0.24675}{2} + \sqrt{0.000197878}}$

$$= 0.5161$$

And
$$B = \sqrt[3]{-\frac{-0.24675}{2} - \sqrt{0.000197878}} = 0.4781$$

Therefore

$$X = A + B = 0.5181 + 0.4781 = 0.9942 \text{ or } 99.42\%$$

3.98 **b.** For a reaction of nth order, $t_{1/2}$ halflife (i.e., at 50% conversion) is given by

$$t_{1/2} = \frac{2^{n-1} - 1}{k(n-1)C_{Ao}^{n-1}}$$

Taking logarithms of both sides, one obtains

$$\ln t_{1/2} = \ln \frac{2^{n-1}}{k(n-1)} - (n-1)\ln C_{Ao}$$

Thus, a plot of ln $t_{1/2}$ vs. ln C_{Ao} gives a straight line of slope $(1 - n)$.

The given data are plotted in Exhibit 3.98.

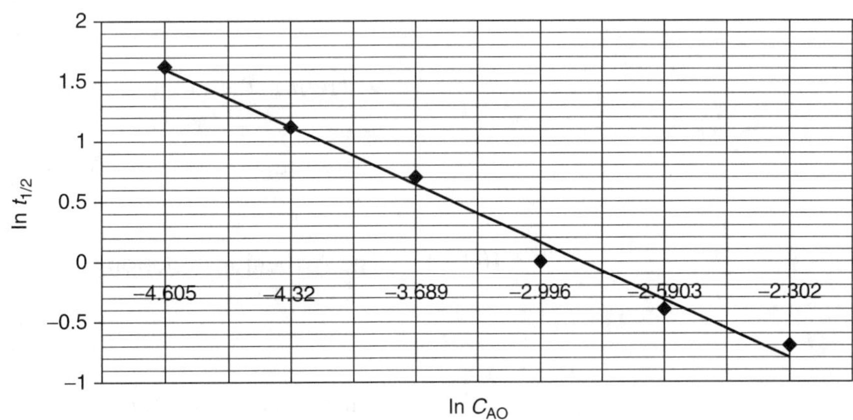

Exhibit 3.98 Plot of ln $t_{1/2}$ vs. ln C_{Ao} for Problem 3.98

$$\text{The slope of the line} = \frac{1.6-(-0.8)}{-4.605-(-2.302)} = 1.0434 \cong -1.0$$

Therefore, $1 - n = -1$ and $n = 2$

3.99 **d.** If all the reactors are maintained at 15°C,

$$k_1 = 0.0806 \text{ min}^{-1}$$

For this case, the exit concentration from the last reactor is given by

$$\frac{C_{Ao}}{C_4} = (1+k\tau)^N = \frac{1.5}{0.075} = 20, \qquad \text{N}\ln(1+k\tau) = \ln(20)$$

$$\text{N} = \frac{\ln 20}{\ln[1+0.0806(7.66)]} = 6.23.$$ (Need more than six but less than seven.)

Since the number of reactors must be a whole number, seven reactors should be used.

3.100 **c.** The conversion for a second order reaction in terms of *Da*, the Damköhler number is given by

$$X = \frac{(1+2Da)-\sqrt{1+4Da}}{2Da}$$

Substituting $Da = 6$, $X = \dfrac{(1+2\times 6)-\sqrt{1+4\times 6}}{2\times 6} = 0.67$

Therefore, percent conversion = 0.67(100) = 67%

3.101 **c.** $$\text{Space time } \tau = \frac{750 \text{ gal}}{7.48 \text{ gal/ft}^3} \times \frac{1}{15 \text{ ft}^3\text{/min}} = 6.69 \text{ min}$$

The Damköhler number, $Da = k\tau = 0.3\,\text{min}^{-1} \times 6.69\,\text{min} = 2.007$

3.102 b. For two reactors of the same volume in series, the conversion is given by

$$X = 1 - \frac{1}{(1+Da)^n}$$

where n is the number of reactors of the same volume in series

Therefore, $$X = 1 - \frac{1}{(1+2.007)^2} = 0.8894 \approx 0.89$$

3.103 b. The Arrhenious relation is

$$k = \alpha e^{-\frac{E}{RT}}$$

If the reaction rate doubles between T_1 and T_2 where $T_2 - T_1 = 10°C$, and k_1 and k_2 are rate constants at T_1 and T_2, one can get the relation

$$\log\left(\frac{k_2}{k_1}\right) = -\frac{E}{2.3R}\left(\frac{1}{T_2} - \frac{1}{T_1}\right)$$

Substituting the values in the above equation one gets

$$\log 2 = -\frac{82{,}324}{2.3(8.314)} \times \left(\frac{1}{T_1+10} - \frac{1}{T_1}\right)$$

which simplifies to $$T_1^2 + 10T_1 - 143{,}014 = 0$$

Solving $$T_1 = \frac{-10 + \sqrt{100 + 4 \times 143{,}014}}{2} = 373.2\,\text{K} \approx 100°\text{C}$$

And then $$T_2 = 373.2 + 10 = 383.2\ \text{K} \approx 110°\text{C}$$

The temperature range between which the rate of the reaction doubles is 100–110°C.

3.104 c. $$C_B = C_{AO}\frac{k_1}{k_2 - k_1}(e^{-k_1 t} - e^{-k_2 t})$$

Differentiating for a maximum, the above equation gives

$$\frac{dC_B}{dt} = C_{AO}\frac{k_1}{k_2 - k_1}(-k_1 e^{-k_1 t} + k_2 e^{-k_2 t}) = 0$$

From which, $$t = \frac{\ln(k_1/k_2)}{k_1 - k_2} = \frac{\ln(0.35/0.13)}{0.35 - 0.13} = 4.55\ \text{h}$$

Then $$C_{B\,\max} = \frac{0.35(5)}{0.13 - 0.35}(e^{-0.35(4.55)} - e^{-0.13(4.55)})$$

$$= 2.79\ \text{lb mol/ft}^3$$

3.105 c. In the case of a single continuous, mixed reactor, differentiating the equation for C_{B1} gives

$$\frac{dC_{B1}}{dt} = (1+k_1\theta)(1+k_2\theta) - \theta[k_1(1+k_2\theta) + k_2(1+k_1\theta)] = 0$$

From which,
$$\theta = \sqrt{\frac{1}{k_1k_2}} = \sqrt{\frac{1}{(0.35)(0.13)}} = 4.7 \text{ h}$$

$$C_{B1} = \frac{(0.35)(5)(4.7)}{[1+(0.35)(4.7)][1+(0.13)(4.7)]} = 1.925 \text{ lb mol/ft}^3$$

3.106 a. For first-order parallel reactions, the rate equations are

$$-\frac{dn_a}{dt} = (k_1 + k_2 + k_3)n_a = kn_a$$

$$\frac{dn_b}{dt} = k_1n_a \qquad \frac{dn_c}{dt} = k_2n_a \qquad \frac{dn_d}{dt} = k_3n_a$$

The solution of the first equation is $n_a = n_{Ao}e^{-kt}$ by direct integration. Substituting this relation into other equations and integrating gives the following:

$$n_b = n_{bo} + \frac{k_1}{k}n_{Ao}(1-e^{-kt})$$

$$n_c = n_{co} + \frac{k_2}{k}n_{Ao}(1-e^{-kt})$$

$$n_d = n_{do} + \frac{k_3}{k}n_{Ao}(1-e^{-kt})$$

After one hour of reaction,

$$n_d = 0 + \frac{0.1}{0.36+0.12+0.1} \times 100[1-e^{(0.36+0.12+0.1)(1)}]$$
$$\doteq 7.59 \text{ g mol of D}$$

3.107 d. For a batch reactor,

$$t = -\int_{C_{Ao}}^{C_A} \frac{dC_A}{-r_A}$$

Thus, the integral

$$-\int_{C_{Ao}}^{C_A} \frac{dC_A}{-r_A}$$

will directly give the batch processing time.

The integral can be obtained by plotting $-1/r_A$ vs. C_A and getting the area under the curve by counting the squares or by the Simpson rule.

To apply the Simpson rule or the trapezoidal rule, it is not absolutely necessary to have the plot.
Obtain the area under the curve as follows:

Apply the trapezoidal rule over $C_A = 0.05$ and $C_A = 0.1$ lb mol/ft^3

$$\text{Area} = \frac{0.025}{2}\left[25 + \frac{25 + 12.3457}{2} + 12.3457\right] = 0.7$$

The area for the concentration range from 0.1 to 0.5 is calculated by Simpson's rule as follows:

$$\text{Area} = \frac{0.1}{3}[1.1765 + 12.3457 + 4(1.8868 + 5.5556) + 2(3.2258)]$$
$$= 1.66 \doteq 1.7$$

Total area = 2.4, which also is the time in hours for each batch.

Therefore, the number of batches = 24/2.4 = 10 batches per day.

3.108 b. For a continuous, mixed flow reactor, space-time is given by

$$\tau = \frac{V}{v_0} = \frac{C_{Ao} - C_A}{-(r_{Af})} = \frac{0.5 - 0.05}{0.04} = 11.25 \text{ h}$$

Therefore, the volume required = 11.25(25) = 281.3 ft^3

3.109 c. For two reactors in series, each with a volume of 50 ft^3, the space-time is given by

$$\tau = \frac{50}{25} = 2 \text{ hours for each vessel}$$

For two reactors, $\tau_2 = n\tau_1 = 2(2) = 4 \text{ hours} = \frac{2}{k}\left[\left(\frac{C_{Ao}}{C_{A2}}\right)^{\frac{1}{2}} - 1\right]$

From Problems 3.107 and 3.108, $C_{A1} = 0.05$ lb mol/ft^3 and $\tau_1 = 11.25$ h

and $$11.25 = \frac{1}{k}\left[\left(\frac{C_{A1}}{C_{Ao}}\right) - 1\right]$$

If we take the ratio of the above two equations, k is eliminated and we can get

$$\frac{4}{11.25} = \frac{2}{1}\frac{\left[\left(\frac{0.5}{C_{A2}}\right) - 1\right]}{\left[\left(\frac{0.5}{0.05}\right) - 1\right]}$$

$$\left[\left(\frac{0.5}{C_{A2}}\right)^{1/2} - 1\right] = \frac{4}{11.25(2)}\left[\left(\frac{0.5}{0.05}\right) - 1\right]$$

From which

$$C_{A2} = 0.074$$

$$\text{Percent conversion} = \frac{0.5 - 0.074}{0.5} \times 100 = 85.2\%$$

3.110 a. For a plug flow reactor,

$$\tau = \frac{V}{v_o} = -\int_{C_{Ao}}^{C_{Af}} \frac{dC_A}{C_{Ao}}$$

The integral is the same value as found in Problem 3.107 for t and is 2.4 hours.

The volume of the plug flow reactor = 25(2.4) = 60 ft^3

PLANT DESIGN AND OPERATIONS

3.111 b. This problem involves the use of equipment cost and Lang installation cost factors.

present estimated cost of a 9019-square-foot unit
= $50,000(9019/3000)^{0.62}$ = \$98,936.

With 3% inflation, the cost of that same unit three years from now
= $1.03^3 \times 98,936$ = \$108,110

Using Lang's installation cost factor (4 for carbon-steel heat exchangers),

approximate installed cost three years from now = 4(108,110)
= \$432,440

3.112 c. Annual cost = $(P - L)[A/P, i, n] + Li + OC$

where
P = initial cost, = \$20,000
L = salvage value, = \$4,000
A = annuity, \$
i = interest rate = 0.1
OC = operating cost per year = \$1000

$$\text{*Capital recovery factor} = (A/P,\ i,\ n) = \frac{i(1+i)^n}{(1+i)^n - 1} = 0.16275$$

*This factor can be calculated or obtained from the tables.

Annual cost = (20,000 – 4000)[0.16275] + 4000(0.1)
+ 1000 = \$4,004

3.113 **a.** For finding out the optimum thickness, it is necessary to plot the total cost against the insulation thickness. The minimum cost point on the total cost curve will give the optimum thickness. To get total costs, read the fixed cost and cost due to heat loss from the curves and, by addition, get the total cost at various thicknesses as follows:

Insulation thickness, in	1	2	3	4	5
Annual fixed cost $	215	355	485	603	735
Cost due to heat loss $	910	610	510	460	425
Total annual costs $	1125	965	995	1063	1160

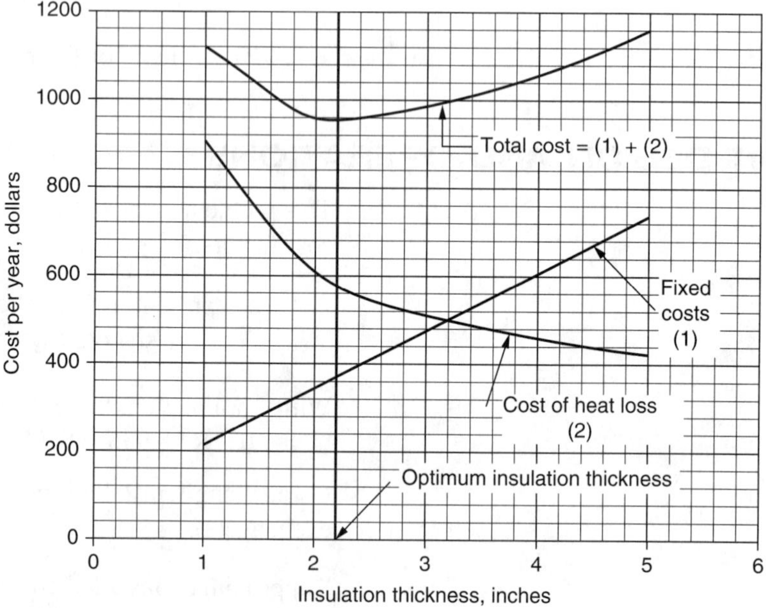

Exhibit 3.113b Determination of optimum insulation thickness for Problem 3.113

Exhibit 3.113b shows these values plotted in the graph provided with the problem (Exhibit 3.113a). On the total cost curve, the minimum cost occurs at insulation thickness of approximately 2.25 inches.

3.114 **c.** The break-even point occurs when

total annual product cost = annual sales.

$$\text{direct production cost/unit} = \frac{350{,}000}{700{,}000/50} = \$25$$

At the break-even point,

$$250{,}000 + 25N = 50N$$

Solving for N,

$$N = 250{,}000/(50 - 25) = 10{,}000 \text{ units/yr}$$

3.115 d. Type 1 settling is free settling and takes place according to Stokes's law. Thus, terminal velocity is given by

$$u_t = \frac{g(\rho_p - \rho_L)d_p^2}{18\mu_L}$$

$$g = 9.81\ \text{m/s}^2 \qquad d_p = 25 \times 10^{-6}\ \text{m}$$
$$\mu = 1.002 \times 10^{-3}\ \text{kg/m} \cdot \text{s}$$

Substitution of values in the equation for u_t

$$u_t = \frac{9.81\frac{\text{m}}{\text{s}^2}[2.5(1000) - 1000]\frac{\text{kg}}{\text{m}^3}(25 \times 10^{-6})^2\text{m}^2}{18 \times (1.002 \times 10^{-3})\frac{\text{kg}}{\text{m} \cdot \text{s}}}$$

$$= 0.00051\ \text{m/s} = 0.00051(3.281)(3600) = 6.024\ \text{ft} \approx 6\ \text{ft}$$

Therefore, the particle will settle 2(6) = 12 ft in 2 hours.

3.116 b. Calculate percentages on the air-free basis.

Total combustibles in air = 2.1 + 0.9 + 0.6 = 3.6%

Percentages of combustibles on air free basis:

Methane % = (2.1/3.6)(100) = 58.33
Hexane % = (0.9/3.6)(100) = 25.00
Ethylene % = (0.6/3.6)(100) = 16.67

Then the lower limit is given by

$$\text{LEL}_{\text{mixture}} = \frac{100}{\frac{\%\ \text{Methane}}{\text{LEL}_{\text{methane}}} + \frac{\%\ \text{Hexane}}{\text{LEL}_{\text{hexane}}} + \frac{\%\ \text{Ethylene}}{\text{LEL}_{\text{ethylene}}}}$$

$$= \frac{100}{\frac{58.33}{5} + \frac{25}{1.1} + \frac{16.67}{27}} = 2.86 \approx 2.9\%$$

3.117 c. Calculate the mole fraction of ethyl alcohol:

	wt %	mol wt	mol	mol Fraction
Ethyl alcohol	90	46	1.9565	0.779
Water	10	18	0.5556	0.221

Assume Raoult's law applies.

Calculate the vapor pressure of pure component ethyl alcohol at its flash point.

$$\log P_i = A - \frac{B}{T + C} = 8.112 - \frac{1592.864}{13 + 226.84}$$

from which $P_i = 29.6$ mm Hg

To obtain this as a partial pressure at another temperature, by Raoult's law,

$$\text{pure component vapor pressure} = (P_i)_T = \frac{p_i}{x_i} = \frac{29.6}{0.779}$$
$$= 38 \text{ mm Hg}$$

Temperature at this vapor pressure is the flash point of 90 wt % ethyl alcohol in water.

Again using Antoine equation,

$$\log 38 = 8.112 - \frac{1592.864}{T + 226.864}$$

Or, $$\frac{1592.864}{T + 226.184} = 8.112 - \log 38 = 6.5322$$

And then, $$T = \frac{1592.864}{6.5322} - 226.184 = 17.7°\text{C}$$

3.118 b. $$\Delta Z = 3000 \text{ ft} = 3000 \times 0.3048 = 914.4 \text{ m}$$

$$T_1 = 100 + 460 = 560°R \qquad T_2 = 155 + 460 = 615°R$$
$$P_{\min} = 3.2 \text{ psia} \qquad P_{\max} = 4.5 \text{ psia}$$

Therefore, the local barometric pressure is

$$P_a = 14.696\left[1 - \frac{0.0065(914.4)}{288}\right]^{5.26433} = 13.2 \text{ psia}$$

The set pressure of the breather vent is

$$P_s = (P_a - P_{\min})\frac{T_2}{T_1} + P_{\max} - P_a$$
$$= (13.2 - 3.2)\frac{615}{560} + 4.5 - 13.2 \approx 2.3 \text{ psig}$$

3.119 a. For horizontal tanks NFPA-30 requires the use of 75% of the total exposed surface irrespective of the vessel elevation (greater or smaller than 30 ft) as for vertical tanks. According to API 2000, 75% of the total surface area or the surface area to a height of 30 ft above grade, whichever is greater, is to be used. In this problem, 75% tank area is greater than the area that would be obtained by using the 30-ft height criterion. Therefore, use 75% surface area.

$$\text{Area of the cylindrical shell} = \pi(D)(L) = \pi(12.5)(40) = 1570.8 \text{ ft}^2$$

$$\text{Area of the ASME heads} = 2(0.918)(12.5^2) = 286.9 \text{ ft}^2$$

$$\text{Wetted area} = 0.75(1570.8 + 286.9) = 1393.3 \text{ ft}^2$$

Heat load per NFPA-30 and API-2000,

$$Q = 963{,}400(F)(A)^{0.338}$$

$F = 1$ since not approved insulation, fire-fighting facility, or drainage. Therefore,

$$= 963{,}400(1)(1393.3)^{0.338} = 11.13 \times 10^6 \text{ Btu/h}$$

3.120 d. Relieving pressure = set pressure + accumulation
= set pressure + 0.21 × set pressure, psig
= 190 + 0.21(190) + 14.7
= 244.6 psia

Calculate the temperature at relieving pressure using the vapor pressure equation as follows:

$$P = (244.6/14.7)(760) = 1264.6 \text{ mm Hg}$$

Substitution in vapor pressure equation,

$$\log 12{,}646 = 6.95805 - \frac{1346.773}{T + 219.63}$$

or

$$\frac{1346.773}{T + 219.63} = 6.95805 - \log 12{,}646 = 2.8651$$

Then

$$T = \frac{1346.773}{2.8651} - 219.63 = 251.9°\text{C} = 525 \text{ K}$$

3.121 c. We need the heat of vaporization at 525 K. Using Watson's relation,

$$\lambda_{T_2} = \lambda_{T_1}\left[\frac{T_C - T}{T_C - T_1}\right]^{0.38}$$

$$\lambda_{T_1} = \frac{14{,}400}{M} = \frac{14{,}400}{92} = 156.52 \text{ Btu/lb}$$

$$\lambda_{T_2} = 156.52\left[\frac{593.9 - 525}{593.9 - 383.6}\right]^{0.38} = 102.4 \text{ Btu/lb}$$

$$\therefore \text{ Relief load} = \frac{11.13 \times 10^6}{102.4} = 108{,}691 \text{ lb/h}$$

3.122 a. The heat capacity of toluene at low pressure and at 251.9°C is given by

$$C_P = 0.09418 + 0.038 \times 10^{-3}T + -0.2786 \times 10^{-6}T^2 + 0.08033 \times 10^{-9}T^3 \frac{\text{kJ}}{\text{mol} \cdot °\text{C}}$$

$$C_P = 0.09418 + 0.038 \times 10^{-3}(252) + -0.2786 \times 10^{-6}(252^2) + 0.08033 \times 10^{-9}(252^3) \frac{\text{kJ}}{\text{mol} \cdot °\text{C}}$$

$$= 0.09418 + 0.009516 - 0.01769 + 0.001286$$

$$= 0.087292 \frac{\text{kJ}}{\text{mol} \cdot {}^\circ\text{C}}$$

$$= 0.087292 \frac{\text{kJ}}{\text{mol} \cdot {}^\circ\text{C}} \times \frac{0.9484 \frac{\text{Btu}}{\text{kJ}}}{\frac{\text{g mol}}{453.6\,\text{g/lb}} \; \frac{{}^\circ\text{C} \times 1.8{}^\circ\text{F}}{{}^\circ\text{C}}}$$

$$= \frac{0.087292 \times 0.9484 \times 453.6}{1.8} \frac{\text{Btu}}{\text{lb mol} \cdot {}^\circ\text{F}}$$

$$= 20.863 \frac{\text{Btu}}{\text{lb mol} \cdot {}^\circ\text{F}}$$

$$\therefore \quad \frac{C_P}{C_V} = \frac{C_P}{C_P - R} = \frac{20.863}{20.863 - 1.99} = 1.1054 = k$$

3.123 a. $k = 1.1054$ (obtained in Problem 3.122)

$$\text{Critical flow pressure } P_{CF} = P_1 \left(\frac{2}{k+1} \right)^{\frac{k}{k+1}} = 244.6 \left(\frac{2}{1.1054 + 1} \right)^{\frac{1.1054}{1.1054+1}}$$

$$= 244.6 \times 0.9734 = 238.1 \text{ psia}$$

3.124 d. The flow will be critical as the pressure downstream of the throat is less than the critical pressure (no back pressure). Critical pressure calculated in Problem 3.123 is 238.1 psia. In this case, the relieving area is given by

$$A = \frac{W}{C K_d P_1 K_b} \sqrt{\frac{TZ}{M}}$$

where

A = relieving area, ft^2
W = relief load, lb/h
C = a coefficient whose value depends on the specific heat ratio k
K_d = coefficient of discharge = 0.975
P_1 = upstream relieving pressure, psia
K_b = capacity correction factor due to back pressure. This applies only to balanced bellows valves only. For conventional valves this factor is 1.
T = relieving temperature of gas of inlet vapor or gas, °R
M = molecular wt of gas or vapor
Z = compressibility factor.
$M = 92$

$$T = 525(1.8) = 945°\text{R}$$

Value of Z

$$P_r = \frac{244.6/14.7}{40.3} = 0.413 \qquad T_r = \frac{525}{593.9} = 0.884$$

From Z chart, $Z \approx 0.81$

Value of C

This can be obtained from Table 9 or Figure 20 API 520, or can be calculated more accurately by using the equation below.

From the table by interpolation,

$$C = 327.5$$

Then
$$A = \frac{108{,}961}{327.5(0.975)(244.6)}\sqrt{\frac{945 \times 0.81}{92}} = 3.816 \text{ in.}^2$$

From the list of standard orifices, an M orifice has an area of 3.6 in.^{2} and an N orifice has a 4.34 in.^{2} area.

Hence, a standard N orifice with area = 4.34 in.^{2} is required.

If the table or chart is not available in the exam, you can use the following relation to calculate C. For the present example, the value of C is calculated below:

$$C = 520\sqrt{k\left(\frac{2}{k+1}\right)^{\frac{k+1}{k-1}}} = 520\sqrt{1.1054\left(\frac{2}{2.1054}\right)^{\frac{2.1054}{0.1054}}} = 327.33$$

3.125 a. Saponification is usually carried out in steel vessels since the solution is alkaline and the corrosion rate is not very high. Cheaper material is therefore selected.

3.126 b. Coupling of two metals far apart in the galvanic series increases the corrosion of the more active metal.

3.127 c. Glass and tantalum are readily attacked by hydrofluoric acid. Therefore, the presence of fluorides rules out selection of these two materials. SS 317L offers good corrosion resistance against wet phosphoric acid, but is not resistant to chloride. Comparing the corrosion resistances of C-276 and Alloy 59 as given, it is apparent that Alloy 59 has a better resistance to corrosion and is likely to last about 13 times longer than C-276 under similar conditions.

Alloy 59 should, therefore, be recommended.

3.128 a. The characteristic equation of the control system can be written as

$$1 + \frac{K_C}{(s+1)(0.5s+1)} \times \frac{3}{s+3} = 0$$

This is simplified to $(s + 1)(0.5s + 1)(s + 3) + 3K_C = 0$

Further simplification gives $\frac{1}{6}S^3 + s^2 + \frac{11}{6}S + 3(1 + K_C) = 0$

We prepare Routh's array as follows

1	$\frac{1}{6}$	$\frac{11}{6}$
2	1	$1 + K_C$
3	$\frac{10-K_C}{6}$	
4	$1 + K_C$	

In order to have all elements of the first row positive and nonzero, $K_C < 10$ since K_C is positive. The system will be stable if K_C is less than 10. However, Routh's test is not adequate to tell about the degree of stability or oscillatory nature of the system. For this, other tests are needed.

3.129 **a.** The transfer function for the set point can be written as

$$C_2(s) = \frac{0.12K_C\left(\frac{1}{\tau s+1}\right)^2}{1+0.12\left(\frac{1}{\tau s+1}\right)^2 e^{-0.5s}} R\ (s)$$

If a unit step change is applied to the set point, the response of the system is given by

$$C_2(s) = \frac{1}{s} \times \frac{0.12K_C\left(\frac{1}{\tau s+1}\right)^2}{1+0.12\left(\frac{1}{\tau s+1}\right)^2 e^{-0.5s}}$$

$$\text{Offset} = R(\infty) - C_2(\infty)$$

$R\ (\infty) = 1$ is *the desired value* since the forcing function is a unit step.

The steady-state value of the response can be obtained by the final value theorem as

$$C_2(\infty) = \lim_{s\to 0}[sf(s)] = \lim_{s\to 0} s\left(\frac{1}{s}\right)\frac{0.12K_C\left(\frac{1}{\tau s+1}\right)^2}{1+0.12K_C\left(\frac{1}{\tau s+1}\right)^2 e^{-0.5s}}$$

$$= \frac{0.12K_C}{1+0.12K_C}$$

$$\text{Therefore the offset} = 1 - \frac{0.12K_C}{1+0.12K_C} = \frac{1}{1+0.12K_C}$$

3.130 b. The operation of a magnetic flow meter is based on Faraday's law of electromagnetic induction. The voltage E induced in a conductor of length D moving through a magnetic field

H is proportional to the velocity V of the conductor. The expression, in mathematical form, is

$E = CHDV$, where C is a dimensional constant.

Thus $$E \propto V$$

SOLUTION AND TOPIC SUMMARY—MASS/ENERGY BALANCES AND THERMODYNAMICS

Problem	Solution	Subtopic
3.1	d	Mass balance, stoichiometry
3.2	b	Mass balance, stoichiometry
3.3	c	Mass balance, excess air
3.4	a	Mass balance, percent conversion
3.5	d	Mass balance, recovery
3.6	a	Mass balance, condensation
3.7	b	Mass balance, ideal gas law
3.8	c	Mass balance, non-ideal behavior of gases
3.9	c	Energy balance
3.10	b	Energy balance
3.11	b	Energy balance, closed system
3.12	d	Energy balance, condensation
3.13	a	Energy balance, heat of vaporization
3.14	b	Energy balance, enthalpy balance
3.15	a	Energy balance, thermodynamics
3.16	c	Thermodynamics, throttling process
3.17	d	Thermodynamics, first and second laws
3.18	b	Thermodynamics, entropy calculation
3.19	c	Energy balance, heat of reaction
3.20	c	Mass balance, HHV and LHV
3.21	b	Energy balance, Watson's equation
3.22	d	Thermodynamics, Watson's equation
3.23	b	Thermodynamics, residual volume
3.24	d	Thermodynamics, entropy, work
3.25	d	Thermodynamics, entropy change
3.26	c	Energy balance, closed system
3.27	a	Thermodynamics, second law
3.28	c	Phase behavior, compressibility
3.29	c	Thermodynamics, second law
3.30	c	Thermodynamics, ideal gas isothermal work

SOLUTION AND TOPIC SUMMARY—FLUID MECHANICS

Problem	Solution	Subtopic
3.31	b	Mechanical Energy balance, Bernoulli equation
3.32	c	Flow through pipe
3.33	b	Measurement of fluids, Pitot tube
3.34	c	Friction in pipes, Fanning friction factor
3.35	b	Friction in pipes, Fanning friction factor
3.36	d	Fluid flow, resistance coefficient
3.37	a	Flow in pipes, Reynolds number
3.38	c	Fluid measurement, orifice meter
3.39	d	Centrifugal pump, TDH, maximum suction pressure
3.40	c	Centrifugal pump, $NPSH_A$
3.41	b	Centrifugal pump, maximum suction pressure
3.42	c	Flow in pipes, equivalent length
3.43	a	Fluid flow through pipes, multiple lines
3.44	b	Compression, adiabatic
3.45	a	Compression, adiabatic, theoretical discharge temperature
3.46	b	Compression, adiabatic, actual discharge temperature with adiabatic efficiency
3.47	d	Isothermal compression, hp determination
3.48	d	Laminar flow, average and point velocities
3.49	d	Flow in partially filled pipes, equivalent diameter
3.50	d	Flow in pipes, effect of increase in flow rate on ΔP

SOLUTION AND TOPIC SUMMARY—HEAT TRANSFER

Problem	Solution	Subtopic
3.51	c	Conduction, resistances in series
3.52	b	Conduction, resistances in series
3.53	b	Convection, dirt factors, overall coefficient
3.54	a	Natural convection outside bundle of tubes
3.55	b	Radiation from pipes
3.56	c	Convection, convection heat transfer coefficient
3.57	b	Convection, Dittus-Boelter equation
3.58	a	Convection, shell side mass velocity
3.59	c	Convection, shell side heat transfer coefficient
3.60	b	Overall coefficient of heat transfer, dirt factor
3.61	d	Convection, Wilson's plot
3.62	a	Convection, Wilson plot, clean tubes
3.63	d	Convection, Wilson plot, h_{di}
3.64	a	Conduction, resistances in series
3.65	c	Conduction, convection, radiation heat loss from pipe
3.66	a	Insulation, resistances in series
3.67	b	Insulation, resistances in series
3.68	c	Radiation, heat exchange by radiation
3.69	c	Radiation, heat loss
3.70	d	Effectiveness factor

SOLUTION AND TOPIC SUMMARY—MASS TRANSFER

Problem	Solution	Subtopic
3.71	a	Diffusion, diffusion of A in non-diffusing B
3.72	b	Diffusion, temperature effect on diffusivity
3.73	b	Diffusion, diffusion through a cylindrical shell
3.74	b	Phase equilibria, Raoult's law
3.75	b	Phase equilibria, *t-x* diagram, azeotropic composition
3.76	c	Phase equilibria, *t-x* diagram, azeotropic composition
3.77	c	Phase equilibria, *t-x* diagram, azeotropic composition
3.78	b	Distillation, McCabe-Thiele method, minimum reflux
3.79	a	Multi-component distillation, short cut method, keys
3.80	b	Fenske equation, minimum number of theoretical stages
3.81	c	Theoretical stages by Erbar-Maddox correlation
3.82	a	Feed plate location, Kirkbride equation
3.83	b	Column operation, % flood
3.84	c	Sieve tray column design, estimation of diameter
3.85	c	Multi-effect evaporation
3.86	c	Multi-effect evaporation
3.87	c	Drying, rotary dryer
3.88	d	Leaching (solid-liquid extraction)
3.89	b	Liquid-liquid extraction
3.90	c	Psychrometry and humidification, use of humidity chart

SOLUTION AND TOPIC SUMMARY—CHEMICAL REACTION ENGINEERING

Problem	Solution	Subtopic
3.91	b	Conversion in a CSTR
3.92	a	Reaction rate constant
3.93	c	Plug flow reactor volume
3.94	a	Free energy of reaction
3.95	b	Heat of reaction as a function of temperature
3.96	c	Equilibrium constant
3.97	a	Equilibrium conversion
3.98	b	Order of reaction
3.99	d	Reactors in series
3.100	c	Damköhler number, conversion
3.101	c	Damköhler number
3.102	b	Reactors in series
3.103	b	Doubling of reaction rate, Arrhenius relation
3.104	c	Consecutive reactions
3.105	c	Consecutive reactions
3.106	a	Parallel reactions
3.107	d	Batch reaction time
3.108	b	Space time
3.109	c	Space time, two reactors, conversion
3.110	a	Heat generation and removal curves.

SOLUTION AND TOPIC SUMMARY—PLANT DESIGN AND OPERATIONS

Problem	Solution	Subtopic
3.111	b	Equipment cost correlations, cost indices
3.112	c	Economic considerations
3.113	a	Operating costs, design optimization
3.114	c	Economic considerations, Breakeven point
3.115	d	Environmental, sedimentation settling
3.116	b	Plant safety, flammability limits
3.117	c	Plant safety, flash point
3.118	b	Equipment design, estimation of set pressure for breather vent
3.119	a	Plant safety, relief load in case of fire in Btu/h
3.120	d	Plant safety, relieving temperature
3.121	c	Plant safety, relief load in lb/h
3.122	a	Plant safety, heat capacity ratio
3.123	a	Plant safety, critical flow pressure
3.124	d	Plant safety, orifice size of safety relief valve
3.125	a	Materials, properties and selection, corrosion considerations
3.126	b	Materials, corrosion considerations
3.127	c	Materials, properties and selection, corrosion considerations
3.128	a	Process control, Routh's test
3.129	a	Process control, offset determination
3.130	b	Process control, sensor

APPENDIX A

Conversion Factors

OUTLINE

LENGTH

	Meter	Centimeter	Millimeter	Foot	Inch
meter	**1**	100	1000	3.281	39.372
centimeter	10^{-2}	**1**	10	3.281×10^{-2}	0.3937
millimeter	0.001	0.1	**1**	3.281×10^{-3}	0.03937
foot	0.3048	30.48	304.8	**1**	12
inch	2.54×10^{-2}	2.54	25.4	0.0833	**1**

1 mile = 5280 ft = 1.61×10^3 meters
1 meter = 10^6 μm (microns) = 10^{10} Å (Angstroms)

VOLUME

	ft^3	in.^{3}	U.S. gallon	m^3	liter
ft^3	**1**	1.728×10^3	7.481	2.831×10^{-2}	28.31
in.^{3}	5.787×10^{-4}	**1**	4.329×10^{-3}	1.639×10^{-5}	1.639×10^{-2}
U.S. gallon	0.1337	2.31×10^2	**1**	3.785×10^{-3}	3.785
m^3	35.31	6.102×10^4	264.2	**1**	1000
liter	3.531×10^{-2}	61.03	0.2642	1.0×10^{-3}	**1**

1 barrel (petroleum) = 42 U.S. gallons

MASS

	pound	grain	oz	gram	kg
pound	**1**	7000	16	453.6	0.4536
grain	1.429×10^{-4}	**1**	2.286×10^{-3}	6.482×10^{-2}	6.482×10^{-5}
oz	0.0625	437.5	**1**	28.35	2.835×10^{-2}
gram	2.2046×10^{-3}	15.4322	3.5274×10^{-2}	**1**	1×10^{-3}
kg	2.2046	1.5432×10^4	35.2736	1000	**1**

1 metric ton = 1000 kg
1 U.S. ton = 2000 lb_m

HEAT, ENERGY, AND WORK

	ft·lb_f	kWh	hp·h	Btu	cal*	Joule (J)
ft·lb_f	**1**	3.766×10^{-7}	5.0505×10^{-2}	1.285×10^{-3}	0.3241	1.356
kWh	2.655×10^6	**1**	1.341	3.4128×10^3	8.6057×10^5	3.6×10^6
hp·h	1.98×10^6	0.7455	**1**	2.545×10^3	6.6142×10^5	2.6845×10^6
Btu	7.7816×10^2	2.930×10^{-4}	3.93×10^{-4}	**1**	2.52×10^2	1.055×10^3
cal*	3.086	1.162×10^{-6}	1.558×10^{-6}	3.97×10^{-3}	**1**	4.184
Joule (J)	0.7376	2.773×10^{-7}	3.725×10^{-2}	9.484×10^{-4}	0.239	**1**

* The thermochemical calorie = 4.184 J; 1 kcal = 4184 J

Energy Conversions

Multiply	By	To Obtain
watt	1	J/s
watt	3.41	Btu/h
W/m^2	0.317	Btu/h·ft^2
W/ m^2·K	0.1761	Btu/h·ft^2·°F
W/(m^2·K/m)	0.58	Btu/(h·ft^2·°F/ft)
kJ	239	cal
kJ	0.9484	Btu

PRESSURE

Pressure Conversions

	mm Hg	in. Hg	bar	atm	kPa	psia
mm Hg	**1**	3.937×10^{-2}	1.333×10^{-3}	1.316×10^{-3}	0.1333	1.934×10^{-2}
in. Hg	25.4	**1**	3.3866×10^{-2}	3.342×10^{-2}	3.3866	0.4912
bar	750.06	29.53	**1**	0.9869	100	14.507
atm	760.0	29.92	1.013	**1**	101.3	14.696
kPa	7.5002	0.2953	1.0×10^{-2}	9.872×10^{-3}	**1**	0.1457
psia	15.701	2.036	6.893×10^{-2}	6.805×10^{-2}	6.893	**1**

1 torr (mm Hg, 0°C) = 1.333×10^2 N/m^2

Pressure Units

1 Pa = 1 N/m^2	1 atm = 1.013 bar
1 bar = 1×10^5 Pa	1 atm = 1.03 kg/cm^2
1 atm = 33.91 ft of water	1 atm = 760 mm Hg
1 atm = 14.696 psia	1 psi = 2.036 in. Hg at 0°C
1 atm = 29.921 in. Hg at 0°C	1 psi = 2.311 ft at 70°F

Values of Standard Atmosphere

atm	1.000
psia	14.696 (14.7 rounded)
ft of water	33.91
in. Hg	29.92
mm Hg	760
kPa (Pa)	101.32 (1.0132×10^5)

POWER CONVERSIONS

	hp	kW	ft·lb$_f$/s	Btu/s	J/s
hp	**1**	0.7457	550	0.7068	7.457×10^2
kW	1.341	**1**	737.56	0.9478	1000.0
ft·lb$_f$/s	1.818×10^{-3}	1.356×10^{-3}	**1**	1.285×10^{-3}	1.356
Btu/s	1.415	1.054	778.16	**1**	1.054×10^3
J/s	1.341×10^{-3}	1.00×10^{-3}	0.7376	9.478×10^{-4}	**1**

VALUES OF IDEAL GAS CONSTANT *R*

Value	Units	
1.987	cal/(g mol·K)	
1.987	Btu/(lb mol·°R)	
10.73	(psia·ft^3)/(lb mol·°R)	
8.314	(kPa·m^3)/(kg mol·K)	= 8.314 J/(g mol·K)
82.06	(cm^3·atm)/(g mol·K)	
0.08206	(L·atm)/(g mol·K)	
21.9	(in. Hg·ft^3)/(lb mol·°R)	
0.7302	(ft^3·atm)/(lb mol·°R)	

VALUES OF CONSTANT g_c IN DIFFERENT UNIT SYSTEMS

	System			
Fundamental Quantity	**SI (International)**	**cgs**	**AES/English Engineering**	**Metric Engineering mks**
Mass (*M*)	kilogram, kg	gram, g	pound mass, lb	kilogram mass, kg
Length (*L*)	meter, m	centimeter, cm	foot, ft	meter, m
Time (*t*)	second, s	second, s	second, s	second, s
Force (*F*)	Newton, N	dyne, dyn	pound force, lb_f	kilogram force, kg_f
g_c	1 kg·m/N·s^2	1 g·cm/dyn·s^2	32.174 lb·ft/lb_f·s^2	9.80665 kg·m/kg_f·s^2

Note that g_c is not the gravitational constant.

VISCOSITY

Units of Absolute Viscosity

	centipoise or 0.01 g/cm·s or 0.01 poise	**slug/ft·s or lb_f·s/ft^2**	**lb/ft·s or poundal·s/ft^2**	**Pa·s or kg/m·s**
centipoise	**1**	2.09×10^{-3}	0.000672	0.001
slug	47,900	**1**	32.2 or *g*	47.9
lb/ft·s	1487	(1/*g*) or 0.0311	**1**	1.487
Pa·s	1000	0.02088	0.672	**1**

Units of Kinematic Viscosity‡

	centistokes	**ft^2/s**	**m^2/s**
centistokes	**1**	1.075×10^{-5}	1×10^{-6}
ft^2/s	92,900	**1**	0.0929
m^2/s	1×10^{6}	10.7643	**1**

*Another conversion factor: 1 centipoise = 2.42 lb/ft·h

‡ν = kinematic viscosity = absolute viscosity/density = μ/ρ.

APPENDIX B

Recommended Reference Data

Some of the problems on the PE exam contain all the data needed to work out the correct solution. Others do not, and you'll want to have appropriate reference data with you in order to solve these.

One of the challenges exam candidates face is deciding which references to bring to the exam. Your stack of books can become bulky and unwieldy, so at a minimum you may want to tab or otherwise designate important sections and pages of key references. Some exam candidates prefer to create a bound reference that compiles information from other sources. The NCEES defines "bound" as books or materials fastened securely in their covers by fasteners that penetrate all papers. Examples are ring binders, spiral binders and notebooks, plastic snap binders, brads, screw posts, and so on.

The best time to gather your reference data is before you start your review work. Prepare two binders, one for use while reviewing for the exam and the other for use during exam.

The following table identifies reference data you will probably want to have on hand during the exam.

Recommended Reference Data	Sources and Remarks
Conversion factors	These are provided as Appendix A of this book. Add any additional factors you think you should have.
Values of ideal gas law constant *R*	These are provided in Appendix A of this book.
Values of standard atmosphere in various units	These are provided in Appendix A of this book.
Standard conditions for the ideal gas	■ Das and Prabhudesai, *Chemical Engineering: License Review,* 3rd ed. (2005), Chicago: Kaplan AEC Education, p. 41 ■ Felder and Rousseau, *Elementary Principles of Chemical Processes,* 3rd ed. (2000), New York: John Wiley & Sons, p. 194
Compressibility charts for low pressure, medium pressure, and high pressure	■ Felder and Rousseau, *Elementary Principles of Chemical Processes,* 3rd ed. (2000), New York: John Wiley & Sons, pp. 208–211 ■ Himmelblau, *Basic Principles and Calculations in Chemical Engineering,* 6th ed. (2004), Upper Saddle River, NJ: Prentice Hall, pp. 285–287 ■ Ludwig, *Applied Process Design for Chemical and Petrochemical Plants* vol. 3, 2nd ed. (1983), Houston, TX: Gulf Publishing, pp. 263–268 These are large charts for different ranges of reduced pressure. You may be able to get these charts from Worthington Corporation, which developed them.

(*Continued*)

(*Continued*)

Recommended Reference Data	Sources and Remarks
Atomic weights and atomic numbers	Felder and Rousseau, *Elementary Principles of Chemical Processes,* 3rd ed. (2000), New York: John Wiley & Sons, facing page back cover
Psychrometric or humidity charts, in both SI and English units	■ Felder and Rousseau, *Elementary Principles of Chemical Processes,* 3rd ed. (2000), New York: John Wiley & Sons, pp. 85–86 ■ Das and Prabhudesai, *Chemical Engineering: License Review,* 3rd ed. (2005), Chicago: Kaplan AEC Education, pp. 517–518 ■ Carrier Corporation, Air Conditioning Division or public relations (best source) ■ Perry and Green, *Chemical Engineers' Handbook,* any edition, Figures 12-1, 12-2, 12-3, and 12-4
Steam tables, in both SI and English units: include properties of saturated steam, temperature and pressure tables, properties of superheated steam	■ Perry and Green, *Chemical Engineers' Handbook,* any edition ■ Felder and Rousseau, *Elementary Principles of Chemical Processes,* 3rd ed. (2000), New York: John Wiley & Sons, pp. 641–642 ■ Himmelblau, *Basic Principles and Calculations in Chemical Engineering,* 6th ed., Upper Saddle River, NJ: Prentice Hall, pp. 642–647
Fanning friction chart with relative roughness as parameter	Perry and Green, *Chemical Engineers' Handbook,* any edition, Figure 5-28 Confusion between Fanning friction factor and Moody-Darcy friction factor should be avoided and appropriate equations used to calculate pressure drop.
Moody-Darcy friction chart	Note that this factor is sometimes referred to as "regular Fanning friction factor." This factor is 4 times the corresponding Fanning friction factor. Best source for this chart is Crane Co. catalogue, Flow of Fluids. It may be found in some texts also. Ludwig, *Design for Chemical and Petrochemical Plants,* vol 1, any edition, Gulf Publishing, Figure 2-2, p. 49 [good size chart, easy to read]
Pipe data, schedule and diameter	Perry and Green, *Chemical Engineers' Handbook* any edition
Drag coefficient for spheres, disks, and cylinders	Perry and Green, *Chemical Engineers' Handbook,* any edition, Figure 5-76
Equations for effectiveness factors of shell-and-tube heat exchangers	Das and Prabhudesai, *Chemical Engineering: License Review,* 3rd ed. (2005), Chicago: Kaplan AEC Education, pp. 261–262
Heat exchanger tube data	■ Perry and Green, *Chemical Engineers' Handbook,* any edition, Table 11-2 ■ Das and Prabhudesai, *Chemical Engineering: License Review,* 3rd ed. (2005), Chicago: Kaplan AEC Education, p. 272
LMTD correction factors for heat exchangers	■ Perry and Green, *Chemical Engineers' Handbook,* any edition, Figure 10-14 (These are small figures and difficult to read.) ■ Kern, *Process Heat transfer,* New York: McGraw-Hill, pp. 828–833 ■ Ludwig, *Applied Process Design for Chemical and Petrochemical Plants,* Vol 3, Gulf publishing Co. pp. 51–56 (good size charts) ■ *Yuba Technical Manual,* Yuba Heat Transfer, Tulsa, Oklahoma
Plot of Erbar-Maddox correlation between reflux ratio and number of stages	■ Das and Prabhudesai, *Chemical Engineering: License Review,* 3rd ed. (2005), Chicago: Kaplan AEC Education, p. 415 ■ Geankoplis, *Transport Processes and Separation Process Principles,* 4th ed. (2004), Upper Saddle River, NJ: Prentice Hall, Figure 11.73, p. 749
Estimation of capacity factor for allowable velocity	Geankoplis, *Transport Processes and Separation Process Principles,* 4th ed. (2004), Upper Saddle River, NJ: Prentice Hall, Figure 11.5-3, p. 730
Pressure drop correlations for random packings	Geankoplis, *Transport Processes and Separation Process Principles,* 4th ed. (2004), Upper Saddle River, NJ: Prentice Hall, Figure 10.6-5, p. 658
Pressure drop correlations for structured packings	■ Geankoplis, *Transport Processes and Separation Process Principles,* 4th ed. (2004), Upper Saddle River, NJ: Prentice Hall, Figure 10.6-6, p. 660 ■ Das and Prabhudesai, *Chemical Engineering: License Review,* 3rd ed. (2005), Chicago: Kaplan Education, Fig 10-20, p. 427; Figure 10.23, p. 429
Generalized pressure drop correlation for packed columns	■ NCEES, *PE Chemical Engineering Sample Questions and Solutions,* (2004) (a good chart with flooding line included.) ■ Perry and Green, *Chemical Engineers' Handbook,* Fig. 18.38

(*Continued*)

Recommended Reference Data	**Sources and Remarks** (*Continued*)
Flooding correlation for packed absorption towers	■ Das and Prabhudesai, *Chemical Engineering License Review,* 3rd ed. (2005), Chicago: Kaplan AEC Education, Figure 12.10, p. 474 ■ *Chem. Eng. Prog. 47*, 423 (1951) ■ Perry and Green, *Chemical Engineers' Handbook*, Figure 18.38, p. 18.22
Distillation and absorption efficiency plots by O'Connell	■ Perry and Green, *Chemical Engineers' Handbook,* Figures 18-23a and b ■ Das and Prabhudesai, *Chemical Engineering: License Review,* 3rd ed. (2005), Chicago: Kaplan AEC Education, p. 676, Figures 18.7 and 18.8
Entrainment correlation	■ Das and Prabhudesai, *Chemical Engineering: License Review,* 3rd ed. (2005), Chicago: Kaplan AEC Education, Figure 18.6, p. 675 ■ Perry and Green, *Chemical Engineers' Handbook,* Figures 18-23a and b
Packing data	■ Das and Prabhudesai, *Chemical Engineering: License Review,* 3rd ed. (2005), Chicago: Kaplan AEC Education, Table 11.3, p. 443 ■ Norton, Koch, Goodloe brochures ■ Geankoplis, *Transport Processes and Separation Process Principles,* 4th ed. (2004), Upper Saddle River, NJ: Prentice Hall, Table 10.6-1, p. 659
API environmental factor table	Das and Prabhudesai, *Chemical Engineering: License Review,* 3rd ed. (2005), Chicago: Kaplan AEC Education, Table 20.6, p. 740
Sizing equations for gas, vapor, or steam	Das and Prabhudesai, *Chemical Engineering: License Review,* 3rd ed. (2005), Chicago: Kaplan AEC Education, Table 20.7, p. 741**
Constant back pressure sizing factor	Das and Prabhudesai, *Chemical Engineering: License Review,* 3rd ed. (2005), Chicago: Kaplan AEC Education, Figure 20.1, p. 742
Back pressure sizing factor	Das and Prabhudesai, *Chemical Engineering: License Review,* 3rd ed. (2005), Chicago: Kaplan AEC Education, Figure 20.2, p. 743**
Sizing equations for gas, vapor, or steam	Das and Prabhudesai, *Chemical Engineering: License Review,* 3rd ed. (2005), Chicago: Kaplan AEC Education, Table 20.7, p. 741**
Sizing equation for liquid	Das and Prabhudesai, *Chemical Engineering: License Review,* 3rd ed. (2005), Chicago: Kaplan AEC Education, Table 20.8, p. 743**
Correction factors	Das and Prabhudesai, *Chemical Engineering: License Review,* 3rd ed. (2005), Chicago: Kaplan AEC Education, pp. 744–745**
Correction factor for two-phase flow	Das and Prabhudesai, *Chemical Engineering: License Review,* 3rd ed. (2005), Chicago: Kaplan AEC Education, Figure 20.6, p. 754**
Thermal venting capacity	Das and Prabhudesai, *Chemical Engineering: License Review,* 3rd ed. (2005), Chicago: Kaplan AEC Education, Table 20.11, p. 755**

** These items can also be obtained from ASME code Pressure vessel design, API 520 etc.

APPENDIX C

Answer Sheets

OUTLINE

MORNING SAMPLE EXAMINATION

Instructions for Morning Session

1. You have 4 hours to work on the morning session. Do not write in this handbook.
2. Answer all 40 questions for a total of 40 answers. There is no penalty for guessing.
3. Work rapidly and use your time effectively. If you do not know the correct answer, skip it and return to it later.
4. Some problems are presented in both metric and English units. Solve either problem.
5. Mark your answer sheet carefully. Fill in the answer space completely. No marks on the workbook will be evaluated. Multiple answers receive no credit. If you make a mistake, erase completely.

Work all 40 problems in 4 hours.

P.E. Chemical Engineering Exam
Morning Session

Ⓐ Ⓑ Ⓒ Fill in the circle that matches your exam booklet.

1.1 Ⓐ Ⓑ Ⓒ Ⓓ	1.11 Ⓐ Ⓑ Ⓒ Ⓓ	1.21 Ⓐ Ⓑ Ⓒ Ⓓ	1.31 Ⓐ Ⓑ Ⓒ Ⓓ
1.2 Ⓐ Ⓑ Ⓒ Ⓓ	1.12 Ⓐ Ⓑ Ⓒ Ⓓ	1.22 Ⓐ Ⓑ Ⓒ Ⓓ	1.32 Ⓐ Ⓑ Ⓒ Ⓓ
1.3 Ⓐ Ⓑ Ⓒ Ⓓ	1.13 Ⓐ Ⓑ Ⓒ Ⓓ	1.23 Ⓐ Ⓑ Ⓒ Ⓓ	1.33 Ⓐ Ⓑ Ⓒ Ⓓ
1.4 Ⓐ Ⓑ Ⓒ Ⓓ	1.14 Ⓐ Ⓑ Ⓒ Ⓓ	1.24 Ⓐ Ⓑ Ⓒ Ⓓ	1.34 Ⓐ Ⓑ Ⓒ Ⓓ
1.5 Ⓐ Ⓑ Ⓒ Ⓓ	1.15 Ⓐ Ⓑ Ⓒ Ⓓ	1.25 Ⓐ Ⓑ Ⓒ Ⓓ	1.35 Ⓐ Ⓑ Ⓒ Ⓓ
1.6 Ⓐ Ⓑ Ⓒ Ⓓ	1.16 Ⓐ Ⓑ Ⓒ Ⓓ	1.26 Ⓐ Ⓑ Ⓒ Ⓓ	1.36 Ⓐ Ⓑ Ⓒ Ⓓ
1.7 Ⓐ Ⓑ Ⓒ Ⓓ	1.17 Ⓐ Ⓑ Ⓒ Ⓓ	1.27 Ⓐ Ⓑ Ⓒ Ⓓ	1.37 Ⓐ Ⓑ Ⓒ Ⓓ
1.8 Ⓐ Ⓑ Ⓒ Ⓓ	1.18 Ⓐ Ⓑ Ⓒ Ⓓ	1.28 Ⓐ Ⓑ Ⓒ Ⓓ	1.38 Ⓐ Ⓑ Ⓒ Ⓓ
1.9 Ⓐ Ⓑ Ⓒ Ⓓ	1.19 Ⓐ Ⓑ Ⓒ Ⓓ	1.29 Ⓐ Ⓑ Ⓒ Ⓓ	1.39 Ⓐ Ⓑ Ⓒ Ⓓ
1.10 Ⓐ Ⓑ Ⓒ Ⓓ	1.20 Ⓐ Ⓑ Ⓒ Ⓓ	1.30 Ⓐ Ⓑ Ⓒ Ⓓ	1.40 Ⓐ Ⓑ Ⓒ Ⓓ

AFTERNOON SAMPLE EXAMINATION

Instructions for Afternoon Session

1. You have 4 hours to work on the afternoon session. Do not write in this handbook.
2. Answer all 40 questions for a total of 40 answers. There is no penalty for guessing.
3. Work rapidly and use your time effectively. If you do not know the correct answer, skip it and return to it later.
4. Some problems are presented in both metric and English units. Solve either problem.
5. Mark your answer sheet carefully. Fill in the answer space completely. No marks on the workbook will be evaluated. Multiple answers receive no credit. If you make a mistake, erase completely.

Work all 40 problems in 4 hours.

P.E. Chemical Engineering Exam
Afternoon Session

Ⓐ Ⓑ Ⓒ Fill in the circle that matches your exam booklet.

2.1 Ⓐ Ⓑ Ⓒ Ⓓ	2.11 Ⓐ Ⓑ Ⓒ Ⓓ	2.21 Ⓐ Ⓑ Ⓒ Ⓓ	2.31 Ⓐ Ⓑ Ⓒ Ⓓ
2.2 Ⓐ Ⓑ Ⓒ Ⓓ	2.12 Ⓐ Ⓑ Ⓒ Ⓓ	2.22 Ⓐ Ⓑ Ⓒ Ⓓ	2.32 Ⓐ Ⓑ Ⓒ Ⓓ
2.3 Ⓐ Ⓑ Ⓒ Ⓓ	2.13 Ⓐ Ⓑ Ⓒ Ⓓ	2.23 Ⓐ Ⓑ Ⓒ Ⓓ	2.33 Ⓐ Ⓑ Ⓒ Ⓓ
2.4 Ⓐ Ⓑ Ⓒ Ⓓ	2.14 Ⓐ Ⓑ Ⓒ Ⓓ	2.24 Ⓐ Ⓑ Ⓒ Ⓓ	2.34 Ⓐ Ⓑ Ⓒ Ⓓ
2.5 Ⓐ Ⓑ Ⓒ Ⓓ	2.15 Ⓐ Ⓑ Ⓒ Ⓓ	2.25 Ⓐ Ⓑ Ⓒ Ⓓ	2.35 Ⓐ Ⓑ Ⓒ Ⓓ
2.6 Ⓐ Ⓑ Ⓒ Ⓓ	2.16 Ⓐ Ⓑ Ⓒ Ⓓ	2.26 Ⓐ Ⓑ Ⓒ Ⓓ	2.36 Ⓐ Ⓑ Ⓒ Ⓓ
2.7 Ⓐ Ⓑ Ⓒ Ⓓ	2.17 Ⓐ Ⓑ Ⓒ Ⓓ	2.27 Ⓐ Ⓑ Ⓒ Ⓓ	2.37 Ⓐ Ⓑ Ⓒ Ⓓ
2.8 Ⓐ Ⓑ Ⓒ Ⓓ	2.18 Ⓐ Ⓑ Ⓒ Ⓓ	2.28 Ⓐ Ⓑ Ⓒ Ⓓ	2.38 Ⓐ Ⓑ Ⓒ Ⓓ
2.9 Ⓐ Ⓑ Ⓒ Ⓓ	2.19 Ⓐ Ⓑ Ⓒ Ⓓ	2.29 Ⓐ Ⓑ Ⓒ Ⓓ	2.39 Ⓐ Ⓑ Ⓒ Ⓓ
2.10 Ⓐ Ⓑ Ⓒ Ⓓ	2.20 Ⓐ Ⓑ Ⓒ Ⓓ	2.30 Ⓐ Ⓑ Ⓒ Ⓓ	2.40 Ⓐ Ⓑ Ⓒ Ⓓ